D0977158

In Praise of *Fire in the Heart*

"This is a story of love, friendship, wildfire, and death written in vivid prose fresh from the fire line. Mary Emerick was a little girl with spindly arms and legs who toughed it out to become a career wildland firefighter, hoping for 'the big one' in the West and filling in the off-season as a 'panther babe' on burns in Florida. There's as much heart as fire in this book. It took a 'big one,' the South Canyon Fire on Storm King Mountain, to bring home the deepest meanings of love, loss, and the bittersweet renewal of life that follows the flames."

—John N. Maclean, author of *Fire on the Mountain: The True Story of the South Canyon Fire*, and other books

"Hour by hour, season by season, fire by fire, readers crisscross the country with Emerick as she rises to the challenges of this difficult life, even as she asks herself whether she can ever stop moving and find another kind of life, where staying might not mean entrapment, where freedom means home."

—Bette Lynch Husted, author of *Above the Clearwater: Living on Stolen Land*

"Imagine facing a wildfire with nothing more than a pulaski and a drip torch. Now, imagine you are a young woman on a crew comprised mainly of athletic young men out to prove something to themselves and to each other. You're carrying a forty-pound pack on your back and slung under the pack is a deployable fire shelter that you may or may not need for your survival. Should the fire jump the line. Should you not be able to reach your safety zone. It will depend on the wind, the relative humidity, and things intangible, like luck. Like fate. Like bravery. *Fire in the Heart* is Mary Emerick's lyrical meditation on one woman's search for identity. A mesmerizing and compelling story of fire and courage and love and transformation."

—Pamela Royes, author of *Temperance Creek: A Memoir*

"Mary Emerick's journey back to nature is one in search of herself. After two decades of fighting wildfires, she came home with the kind of stories we all need to hear; stories that help us understand that at times, we can be brave, we can be strong, we can be fully human."

—Murry A. Taylor, author of *Jumping Fire: A Smokejumper's Memoir of Fighting Wildfire*

"Beautiful. Riveting. Satisfying. An honor to read."

—Ellen Airgood, author of *South of Superior*

"Emerick's story unfolds much like a fire. There are quiet moments at dusk, staring off into the distance, mesmerized by the particular beauty of fire. How landscapes and people take hold of our lives and change us. How wildfire and fighting fire regenerate. And frantic moments of trying to stay one step ahead: of catastrophe, of personal transformation, of grief. To read *Fire in the Heart* is to become part of the link in the chain, to find your place on the fire line, to understand more deeply what it means to be a wildlands firefighter, and to keep one foot always in the black."

—Cameron Keller Scott, author of
The Book of Cold Mountain,
winner of the Blue Light Press Poetry Prize

"Masterful! A beautiful, gripping exploration of Mary Emerick's twenty-five-year journey through the insular, dangerous, hyper-masculine world of wildfire. In riveting prose, Emerick writes of what she gave up—and what she gained—by choosing a life as a wildland firefighter. *Fire in the Heart* is a thoughtful meditation on the impact of wildland fire suppression on the American landscape, and on one woman's heart. I love this book!"

—Mary Pauline Lowry, author of *Wildfire: A Novel*

"A tough and tender tale of human bonds forged through fire. Riveting."

—Elizabeth Enslin, author of *While the Gods Were Sleeping: A Journey Through Love and Rebellion in Nepal*

Fire *in the* Heart

Also by Mary Emerick

The Geography of Water: A Novel
(University of Alaska Press, 2015)

Fire *in the* Heart

A MEMOIR OF FRIENDSHIP, LOSS, AND WILDFIRE

MARY EMERICK

Arcade Publishing • New York

Arcade Publishing books may be purchased in bulk at special discounts for sales promotion, corporate gifts, fund-raising, or educational purposes. Special editions can also be created to specifications. For details, contact the Special Sales Department, Arcade Publishing, 307 West 36th Street, 11th Floor, New York, NY 10018 or arcade@skyhorsepublishing.com.

Visit our website at www.arcadepub.com.

10 9 8 7 6 5 4 3 2 1

Library of Congress Cataloging-in-Publication Data is available on file.

Cover design by Erin Seaward-Hiatt
Cover photo credit iStockphoto

Print ISBN: 978-1-62872-843-9
Ebook ISBN: 978-1-62872-847-7

Printed in the United States

Table of Contents

Through our great good fortune, in our youth our hearts were touched with fire. It was given to us to learn at the outset that life is a profound and passionate thing.

—Oliver Wendell Holmes Jr.

Holding the Line

I had been fighting fire for so long that I was not even sure what day it was. In the last two weeks the days had blurred together in a constant waking dream of smoke and fatigue. Roll out of the sleeping bag, pull on stiff leather boots, grab my pack and tool, dig fireline for sixteen hours, fall into bed, clothes still on. Repeat.

The state of my flame-resistant Nomex shirt might be a clue. I must have been wearing it for at least five days for it to smell this bad. My Kevlar pants were worse, stiffly crusted with spilled saw fuel. I thought I remembered taking a shower two days ago, though my legs were permanently stained black from walking through knee-deep ash.

My long hair was knotted into dreadlocks under my sky-blue hard hat; my lips scabbed from sun and wind. I did not look or feel like a woman anymore. I was not anything substantial, just a constant motion. I only bent with the pulaski in a kind of endless dance. Scrape the duff down to mineral soil. Take another step. Ignore the sweat that trickled down my neck and between my breasts. Shove everything else—hunger, thirst, regret, fear—deep beneath, in some other place.

The sound of deep fire coughs echoed down the line. We had all sucked in enough smoke to equal two packs today. There were no masks light enough to wear and still do this job. We did it half-assed instead, pulling bandannas over our noses and mouths. The smoke filtered in anyway. Weeks after I left here I knew that the tightness in my chest would linger.

The rest of the twenty-person fire crew were falling into the usual grooves, the kind that you ground into after a few days on the same fire. At first, everyone had kept their mouths shut and their tools flying, but after a few tough shifts, I could size up the crew pretty well. I knew who the slackers were, and the freelancers, and the good ones. There were those who could save your butt if things went south, and others who would fall apart, lose it, and get burned up.

I couldn't think about that tonight. Instead I kept an eye on the crew, because invariably they were doing something they should not. "Look into the green!" I yelled down the line. The rookies were making the typical mistake of staring, mesmerized, at the fire itself. It was an impressive sight as it jumped into the tops of black spruce and sizzled in the oven-dry needles. But where we really needed to be looking was in the green, unburnt section, our backs to the fire. This was where spot fires could blossom, caused by unseen sparks tossed across by wind. Firebrands, they were called, and the analysts in camp carefully concocted predictions of ignition in terms of percentages.

The scale went up to one hundred percent, meaning that if an ember were to fall on unshaded vegetation, there was a hundred percent chance of a spot fire beginning there.

Today was more like fifty percent, a long night stretching ahead of us without the adrenaline rush of spot fires. We had pulled the night shift again, or more accurately the dim shift, since this far north in Alaska the sun never really set, just sort of puttered around near the edge of the horizon. There was already a snap in the air, which did not bode well for staying warm. Our initial attack backpacks were already stuffed to the brim with extra food and water, so it would be a cold night without the sweatshirt left behind at camp. A loud thrashing in the brush could be a moose; one had been spotted on the fireline.

We were spaced out evenly along the line, the next person a hundred feet down the hill, so we could only shout to each other. Already there was mutiny in the ranks, evidenced by sounds of loud unnecessary chopping. Several green aspens already lay skinless and bleeding on the green side of the line. The sawyers were knocking down trees for the hell of it, creating their own clear cut. Each thump was greeted with a rebel yell, as if they had conquered something bigger than themselves.

"Hey! Stop sport falling!" I yelled.

"Yeah, whatever," Dan, the ringleader, yelled back.

I sighed.

Years after we were gone, the scars we left would still be here—our three-foot line, snaking through the permafrost, the ground laid bare, unzipped to its core, the stumps, the tracks of dozers. I thought of all the other fires, in Florida, Wyoming, Colorado—long swathes of forest changed forever, not only by the fire but by our stopping it. The cut trees, the brush piles, the thin red line of fire retardant.

My mind wandered. How many times had I stood just like this, shivering a little bit in my thin Nomex shirt, carrying a pulaski, holding the line? There had been nights of spitting rain, huddling under flimsy space blankets. Nights curled around warm stumps, praying for the sun to poke out from the other side of the mountains. Nights spent hauling water up the line, each of us sunk in our own private misery, icy water from leaky bladder bags snaking down our backs.

There had been a hundred other nights like these, both on the fireline and on a trail, countless hungry, freezing, desperate nights that were sometimes easy to forget. When was I going to call it quits, say enough was enough? I was forty years old. Shouldn't I have had kids, taken them to soccer practice, become another, different woman with shiny hair sweeping my cheeks at a careful angle, fingernails the color of tangerine, everything perfect, not used up, stained, broken? What kept me here, what was this thing that grabbed me and wouldn't let go?

I looked down the line at the rest of the crew. There was a row of hard hats as far as I could see, stretching down the line and around the corner. Some leaned on tools, their chins resting on the handles, staring into space. Others fiddled with Meals, Ready to Eat; packets full of lumpy, mysterious stews and pound cakes the consistency of sawdust. All of us were here because we were fascinated by fire. Some said it was for the money, a little extra to see them through the lean winters. But press each of them a little bit, and it was something other than money that kept us out here season after season. One thing was for sure: you either loved it or you hated it. There was no in between.

A rookie named Robin hiked back up the hill to my position, a halo of fuzzy blonde hair escaping the confines of her hard hat. "Hey, what are we supposed to be doing?" she asked, planting her shovel in the skinny line that separated burnt ground from green.

It was a clear violation of the rules for us to be clumped up like this, but I welcomed the distraction. This was her first fire and she was not sure yet whether she liked it. Coming from an academic world where everything made sense, this pointless standing around bothered her.

"Why are we so far from our safety zone?" she asked.

We watched the main fire back down the hill through a cluster of hardwoods. Going against the wind, it was slower, almost casual, the flames here only waist-high. It seemed almost benign, inconsequential. It was hard to believe that this tame fire was related to the creature that roared in the canopy of the black spruce or the same thing I saw tearing through the prairies in Florida. The same thing that killed. It would be easy to turn my back, to discount it, if I didn't know better.

Robin echoed my thoughts. "The hardwoods are burning. Where are we supposed to go if we have to get out?"

I didn't know what to tell her; what we were doing was miles from fire school, where situations were put in simplistic terms. It was black and white. Never do this. Always do that. Out here it got messy and complicated. Sometimes, something you had taken for truth turned around and bit you in the butt.

"I think we'll be okay," I said insincerely, though in all honesty I wasn't sure. "See how cold it's getting? The fire should lie down soon." She nodded, unconvinced.

Laughter broke out to our south. Somewhere in the black, a few brave souls decided to huddle around a burning stump, risking the Division Supervisor's wrath. The Kalskag crew, farther down the hill, had pulled out their coffee pot a long time ago. They put on black crew sweatshirts and talked softly. When they didn't want us to understand, they spoke in their native Yupik.

Robin and I looked at each other. "Screw it. I haven't seen the Div Sup in hours. Let's go over," I said. "What are they going to do? Fire us? Send us home?" She laughed.

Robin and I cut across an area that had been nuked, everything green and alive gone. This was what we called "good black." The ground was crunchy under our boots, the consistency of burnt potato chips. The heart of the fire had passed through here, and there was no chance of any reburn. It was a safe place to be. It was what we looked for on every fire, a solid carpet of black, a place where we could let our guard down for a little bit.

The rest of the crew shifted to let us in. We leaned against our packaged fire shelters, the flame-resistant tents that were supposed to save our lives. I knew by now this was not always true. Not everyone who went in a shelter came back out alive. They made a comfortable backrest, though, even though we had been warned that this practice damaged them. We didn't plan on using them anyway. That was another deception we told ourselves out here.

That'll never happen to us.

The crew had been telling fire stories, each one more badass than the one before. Five pairs of eyes turned to me.

"You're an old fire dog. You've been doing this for how long, since the eighties?" a rookie wanted to know. He was eighteen.

I had been fighting fire longer than he had been alive. He couldn't imagine rebellious knees, unstable from years of holding the line. He didn't know how hard it was to dig a little deeper each year to stay ahead on the hikes. He did not have the memories, polished stones tumbling around in my head: Florida to Alaska; grass, desert, and mountain; all different but all somehow the same.

He stirred the fire with a stick. "Were you at Yellowstone? South Canyon? You must have some stories."

They waited. Time was irrelevant on the night shift. Each minute contained an hour. The fire crackled to itself. We were miles away from a town. A dim light spanned the horizon, the remains of sunlight. At another spot a couple of others, following our lead, quit chopping and hunkered down over their own warming fire. I could see Dan's wiry silhouette outlined in the flames as he threw logs onto the fire, building it up way past what was acceptable. I could hear him winding up one of his interminable stories. "The time that I was in jail," he began. Sensing a captive audience, he warmed to a tale that may or may not have been true; none of us could ever tell. I gritted my teeth; Dan had been on my last nerve ever since the beginning, galloping downhill into the fire without scouting, standing too close to the sawyers as they felled danger trees. Unconvinced of his mortality, he sizzled with energy that he seemed unable to contain. Nothing I said convinced him to rein it in.

Under cover of the night I could talk about the years of fighting fire. Kentucky, Wyoming, Texas, and all the places in between. I could talk about saving houses and back burning and mopping up. I could talk about sleeping on a mountain's shoulder somewhere near a nameless lake. But there was still one fire I couldn't talk about. One fire that would not let go of me. I was afraid I would forever be in its grasp.

The kid asked the question again, the one that I had been avoiding for ten years. "Were you at South Canyon?" he asked. "You know the one where thirteen people died?"

Fourteen, I thought. There were fourteen. "No," I said. "I wasn't there."

It was true that I wasn't there. But what was also true was that I carried that fire with me everywhere.

There were still hours to go, hours that would stretch into days on this fire. Despite our best efforts, all of our saw lines and hose lays and standing here all night with our tools, the fire would go nuclear, crossing the Little Chena. The rains would come too, the one thing that really could put the fire out. In the spring it would happen all over again, an endless loop.

As I sat there it came to me. I had been running away all my life. It was, finally, time to stop.

The fireline was just another way to run. I didn't know if I would ever see myself the way I wanted to be. I couldn't outrun the girl I used to be; I'd tried that already. I had not drowned her in the rivers we crossed; I had not lost her in the canyons. She would always be a part of me, like it or not.

I thought of the other women I knew who faced no such dilemma. Not firefighters, they never had to grab two forty-pound cubitainers of water and a pulaski and haul them uphill. Nobody had ever yelled at them to close the gap because they were walking too slowly in the single-file line. They were content to let men help them. They had never felt like they needed to prove anything.

For so long my identity had been as a firefighter. I had spent years building this persona, years of trying to crank out enough pull-ups, my hands sweating on the cold metal bar, years of gritting my teeth to keep up on the hikes. Only I knew that it now felt like a carefully orchestrated act. The person that others saw was someone I had created, layers of thick skin enveloping the softer girl who was hidden underneath. The one who would sometimes have liked to lean on someone. The one who would sometimes have liked to stop.

The men on the fire crew offered few answers. They seemed content to live this life until they were halted by their failing bodies. They seemed equally able to put the past behind them, whether that was their younger self or a cross on a burnt hillside. Their conversation was an equal mix of trash and longing, stories of girlfriends who didn't understand why they couldn't stay home. The women on the crew earned their respect by carrying loads without complaint, but they didn't want to marry us. Their women were lipsticked and glossy, not firefighters.

Who would I be without fire?

I recognized my own younger self in every set of bright eyes around the warming fire, every flash of white teeth against a grimy face. There was probably more to all of them that lay beneath, just like there was with me. Probably each of them had something that split them in two, something they were trying to cover up by diffidently spitting Red Man into the fire or wielding the chainsaw in a shower of falling sparks. I was not the only one who wondered whether to go or stay. I was not the only one haunted by a fire from long ago.

As I sat there, thinking of the right words to speak, I knew deep in my heart that it was time to try to stay home. It was time to hang up my fire boots for good. Turn over my gear to the next girl, the one with bright eyes and a long ponytail, the one who did not have as much weight to carry. It was time to go back to South Canyon, to climb Storm King Mountain, to finally hear what the mountain had to tell me.

Burning the Prairies

One of the best memories I had of the three of us was the last time we burned the prairies together. I could still recall everything about that day. I could feel the weight of the squat steel canister I clutched in one gloved hand, smell the combined kiss of diesel and unleaded as I lit a match to the wick. I could see the three of us walking parallel, lines of fire behind us, merging together into one blaze.

The effortless Florida dusk dropped like a heavy stage curtain and a distant glow marked the path of the main fire as it tore toward us across a savanna of sawgrass.

The smoke column rose thousands of feet into the air, high enough into the atmosphere that it created its own weather. Today it brewed up a thunderstorm: lightning forked out of the smoke and a few drops of rain sprinkled onto our blue hard hats under an otherwise cloudless sky. Tourists headed up Alligator Alley, the main highway belting the state, pulled over to take pictures, running with cameras and flip-flops along the safety fence.

Roger, Jen, and I were on foot, carrying our long-handled flappers and drip torches as we finished up our burnout, the blackened edge that would ring the prairie. As we walked, fat eastern diamondbacks wriggled out of our backfire and across the trail.

"Watch out for burning bunnies," Roger called. It was true: an animal, fleeing our fire, could spread flames across into the unburnt "green" side and trap us.

Sometimes the snakes, inexplicably, turned and wriggled back toward the fire and certain death. We had seen it before, an entire field covered with dead snakes. Even though none of us really liked snakes, we tried to save them with shovels, herding them out into the unburnt neighboring unit. They resisted, striking at the metal blade and slipping past us back into the heart of the fire.

What lay under our boots, hidden by a thin layer of soil, was an immense field of limestone, tipped so that underground water flowed over it toward the Gulf. In places the acidic water had eaten through the soft rock, making grooves where trees had taken over from the prairies. Here, mere inches determined what grew. There were the lower-lying cypress strands, awash with black water, and the slightly higher hardwoods, oaks and pond apples and strangler figs in mounds called hammocks. Often, walking through those, we found the bones of small animals that had died there, stranded during the wet season. They starved, running out of food on an island refuge.

Another few inches and we encountered pine stands with a soft grass understory. Drop low and we were in the pancake-flat prairies, dotted with solution holes as the limestone eroded away. Walking through the prairie, hearing its whisper against my thighs, sometimes made me feel like I was being swallowed by grass. It could grow so tall it was over my head and I would become lost momentarily, adrift in a pool of grass, like a swimmer on her back. I imagined backstroking all the way to the Gulf in the summer wet season, my long hair mingling with the grass until we merged into one being.

What the fire did here depended on a delicate balance of moisture and ignition. If it was too wet, the air musky and swollen, flames flickered out. If it got dry, below thirty percent humidity, fires ripped across the prairies and thundered through the cabbage palms. Even the green, live vegetation burned; saturated with volatile oils, it sizzled and torched high into the sky. This was a country built to burn.

Our progress that day was slow and even, marked by triple lines of smoke. With our drip torches, our task was to burn out all the fuel in front of the main fire. As we walked we kept an eye on the wind, because a shift could mean flames shooting past us and our fireline. We kept an eye on each other too.

The swamp buggy, an oversized jeep built just for the Florida swamps, lumbered behind us, its slip-on pump clattering in readiness. The two Mikes, one on loan from the state park and the other a seasonal worker on the wildlife refuge crew like Jen and Roger and me, had drawn the short straws today and sat resigned on bucket seats, Nomex shrouds cinched tight around their faces. The buggy crew choked the most smoke and was doomed to crawl along behind the action, either dousing spot fires across the line or cooling down burning snags. It was a boring job at best and the Mikes faced a long evening of laborious laps around the fireline while the rest of us skipped off to clean up. We would be in shorts and bare feet while they were still peering through the smoky circle of headlights.

Jed, our boss, had already called the shots with assignments, and fair or not, we took them. No matter that the Mikes had been stuck on the buggy for the last two burns, or that someone else got the freedom of rocketing around on the ATV again. What Jed said went. None of us would think to argue.

Close-cropped peppery hair, steel-cut eyes—Jed set high standards for us and we worked hard to meet them. He seemed ages older than me although in actual years it was less than ten. Fire did that to you. It put creases in your face from squinting across canyons to size up the flames. It settled in deep on your shoulders, responsibility for the lives of others resting heavy.

Our safety briefing this morning was sprinkled with his colorful Arkansas-born sayings. "Roger, you run the lighting show. Take one or two burners, whatever melts your butter. Keep an eye on the east line. When this gets romping and stomping, I don't want everyone scattered from here to breakfast."

He adjusted his floppy hat on his head and fixed us with a hard stare. "With this wind, we'll be busier than long-tailed cats in a room full of rocking chairs." The message was clear: don't screw up.

This was the thing about prairies: they were easy to burn, as easy as laying down a match. We burned them hot, as hot as they would get, lighting them so the wind chased the flames. The fire rolled across the grass faster than we could run, in an undulating carpet of red and gold. It burned so hot that even the water flowing under the sawgrass looked like it was on fire.

The main fire was lit from the helicopter by Steve, our foreman. From the air, he burned the interior that we could not safely reach; from the ground, we burned the fringes, creating a black edge. That ensured two things: fire would not cross, or slop over, to the adjacent unit, and the heat from the main fire would suck ours toward it, preventing a wall of flame heading in our direction.

Wind was the main ingredient and in places other than prairies it was easy to get wrong. Too light and the fire might fizzle, not meeting the wildlife refuge objectives. Too strong and the fire would rip through the pines, killing them. Direction was important too. Head fires ran with the wind, the hottest, most dangerous fire there was. This was all right for the resilient prairies but deadly for the oaks and the pines. A flanking fire, one that slunk perpendicular to the wind, or a backing fire, moving slowly against it, was better. All the ingredients had to come together to pull off a successful burn.

I had come to Florida wanting to fight fire, but we were starting more fires than we put out. We were using fire to shape the land, setting the prairie on fire on purpose. The stated reason was to reintroduce fire into the landscape, as if it were a stranger long gone from these parts. Fire, meet prairie. Prairie, meet fire. This part of Florida, south of Tampa and north of the Keys, was so irrevocably messed up—canals diverting the lifeblood that used to flow to the Everglades, concrete covering over what once was slash pine forest—that it was uncertain if our small effort made a difference.

Adding fire back in where it used to belong, Jed told us, put the missing piece back in the puzzle that was the

Everglades. The plants and animals we saw here had grown up with fire, learning how to tolerate it in several ways. Take the slash pine, he said, the main conifer on the wildlife refuge. The young trees were killed easily by fire, but by the age of ten, they became more tolerant of it, developing a scaly bark that flaked off, shedding embers. Fire, passing through, killed off their enemies the hardwoods and laid the soil bare, helping new pines generate. This also cleared out the undergrowth, all the vines and ivy and thick palmetto that crept in without fire's broom. Deer couldn't pass through the fortress of all this vegetation and neither could their main predator, the endangered Florida panther. Fire was the thread that kept everything in balance. Pull out the thread, and it all unraveled.

I had never looked at a forest before and picked it apart, seeing what belonged and what didn't. There were ghosts of the missing: bees that used to pollinate the fleshy lips of the cigar orchids; swallow-tailed kites, rare now. There were the invading armies: Brazilian pepper with its blood-colored berries and glossy green leaves; thirsty meleleuca, once seeded from airplanes in an effort to dry up the swamp. Sometimes it seemed like everything was on the edge, and the golf courses and condos would swallow up our refuge in one hungry gulp.

We kept at it, though, setting the prairies alight once every three or four years. This was what was called a "prescribed fire," as though we were prairie surgeons. Fire in small doses, Jed had told us in one of his impromptu fireline lectures, renewed the grasses and cleared out the choked brush in the forest. Deer could slip through the clearings, followed by the endangered Florida panthers we were all desperately trying to save.

A fat moon rose as Roger headed into an unburnt patch with a drip torch, bringing a line of fire with him as he walked. "I'm being followed by a moon fella," he sang loudly and off-key, convinced that those were the correct lyrics to the Cat Stevens song. Jen and I, waiting with our torches, collapsed in laughter.

"*Moon fella?*" we screamed, delighted, and he turned to look at us, fire framing his lanky silhouette. "What?" he called, not understanding. With bangs hanging in his face under his hard hat, he looked about fifteen. Then he was swallowed by the tall grass. We marked his progress by the line of smoke rolling out from the prairie and took up our positions, Jen halfway between Roger and me.

The Jet Ranger helicopter clattered near us, finishing up strips in the interior. Steve had the best job of us all, hanging out from the back seat of the helicopter, one foot on the skid as he fed little white spheres we called ping pong balls into the aerial dispenser. Riding in a helicopter was as close as you could get to flying yourself. Up there he had a view of the entire fire that we didn't; he could see the blackened edges we walked and the fiery heart of the interior flames. From up there, we looked like small puzzle pieces that Jed, sitting in the front seat, rearranged depending on how the burn was going.

Looking up, I saw the balls fall like hail onto the prairie below. With a skim of potassium permanganate inside, and injected at the last minute with glycol, they ignited in twenty seconds. This put more heat on the ground and herded the main fire where we wanted it to go. When he got close, Steve amused himself by throwing a few uninjected balls at our feet, watching us scatter.

"Hoot!" Roger yelped from the woods.

"Hoot!" Jen and I chorused back. It was still impossible to see him and it was important to keep our line slightly staggered so that I, the last burner on the trail, hung back behind him. Otherwise my fire and Jen's could flank together and move toward him before he could run out of the way. This had happened to people before. They had died this way, from friendly fire.

Tracks clanked as Bill, our heavy equipment operator, walked the dozer slowly up to the next corner. He puffed a cigarette, looking contemplative. "Put a fork in it, it's done," he said. Last week a psychic told him he would meet a beautiful dark-haired woman and retire from his job with a disability. Since then he had been mentioning his aching back more frequently. The woman had yet to make an appearance.

Burning season was just about through. There might be one, two more weeks. Maybe a month, if it rained some more. Already we could see changes in the weather, the wind switching more often from the southeast, cold fronts no longer able to push down this far. Summer was on the way with its higher humidity and daily thunderstorms. We could burn up to about seventy percent relative humidity, but after that nothing lit well. Instead we ended up with patchy, dirty burns, black splotches in the green.

Soon those of us on the seasonal fire crew would pack up our trucks and head west—Roger back to McCall to work as a smokejumper for the summer season, Jen to build trails at an island park in Lake Superior, and me to work as a Forest Service ranger in the Idaho wilderness. Both Mikes were locals, our Mike picking up summer work in the off time, and Fakahatchee Mike, who was just on loan to us for some burns, went back to the state park we nicknamed him after. The rest stayed to slog through the rainy hurricane season.

Steve, Bill, and Jed held down the fort, overhauling the equipment and catching up on paperwork. If they were lucky and it was a good western fire season, they could get called up to go out there too.

The three of us who left never thought too much about it, but we all assumed that each season would be like the last. There would always be prairies to burn, and it would always be us three out here burning them.

The grass parted and Roger emerged, blowing out his torch. "Tied in," he said. We stood and observed our handiwork with pride. Our flanking fire joined together and swept like an ocean wave to meet the main fire, colliding in a confused rush of flame. The fire jumped and whirled a hundred feet high in a symphony of crackling sound.

A big fire front did not have just one noise. There was the high-pitched whine as resins and oils caught, the crackle-pop of crunchy dry leaves. There was the thunder of trees bursting apart and falling and the sizzle of sap as flames climbed high, wrapping around the trunks.

Way out in the prairie where the fire had nearly burned itself out, a dense glowing carpet of embers like fallen stars spread out for a half mile. An orange glow filled the slice of sky that was visible over the tops of the trees. Its light was reflected on the faces of my friends next to me.

Watching this, I swallowed hard, a lump in my throat that I couldn't quite explain. This one moment was so perfect it felt like love and I wanted to wrap my arms around all of it, to stop the night from turning and the fire from going out. For a moment I couldn't move, my boots firmly rooted to the damp soil beneath. The love I felt right now, for what we were doing here and for the two friends who stood here with me, was what I had been looking for all of my life.

In the distance I saw the buggy slogging toward us, back from its mission of spraying down the billboards by the highway to save them from fire. "We'd better do something even if it's wrong," Roger concluded. He was right: there was always something to do. We couldn't stand here all night, even though I wanted to let the dark velvet absorb into my skin and watch the fire for as long as I could.

Roger and Jen began to move and I walked with them. We walked through the burnt grass, flattened and black under our boots. It seemed that nothing could survive this, but in only a few days, small juicy shoots would poke out—deer candy. This land was more resilient than it seemed, but in other places around here it had taken a beating. Canals had been dug, diverting the water that was the lifeblood of this grassland. Development sprawled over the places where we had fought fire only a few seasons before, the cabbage palms and slash pines taken out. The panthers we were trying to save hovered around thirty adults, decimated by mercury and habitat fragmentation. Our wildlife refuge was a small island in a sea of development. It was not enough.

Though we could not know it then, this was the last time all three of us would burn a prairie together. That evening walk through the tall grass, all three of us in a staggered parallel line, separate rows of fire following each of us, was the last walk we would ever take together. As we walked, the three lines merged together, none distinct and separate anymore, but indistinguishable, gone forever.

In less than six months, one of us would be dead.

The Accidental Firefighter

In the beginning I was an accidental firefighter, pressed into service because there was a state of emergency in the Northwest. Four months earlier I had floated across the country like a dandelion seed to land at Olympic National Park for the summer. Most of my friends at home were locked in, framing their college degrees, settling into marriage and real jobs. Some of us, though, took different routes, striking out in beat-up cars long past their prime, long-haired girls with unexpressed dreams. Most of them came back home. "You'll be back," they said as I drove away. I was determined to beat the odds.

There had to be something more, I thought—a country I had read about in books, a country that could change a woman. When my friends came back, sometimes with a new puppy or a new man in tow, they seemed exotic and new, touched by distant mountains. After a time, though, they settled back into the familiar landscape of well-trodden paths: the bar on Saturday nights, wearing too much blue eye shadow, a dismal job at a desk, an inescapable history.

I had grown up loving the woods, and seasonal work at a national park was easy to get, even for someone with an English degree like me. You just had to accept five dollars an hour for your pay, drink cheap whiskey passed hand to hand on bunkhouse porches, and believe promises from sweet-talking men that did not last more than a summer. I had never had an adventure of my own, growing up sheltered and soft, and I craved those late nights surrounded by people who had been everywhere, who didn't think past the next ninety days. Unlike where I grew up, your past did not follow you.

The park was immense and sprawling, with moss dripping from ancient trees and peaks that held snow year-round. I was from an iron-flat state punctuated with bumps we called mountains. The woods there were fenced off and private, slender second growth logged out long before, woods with no mystery. Out here, the whole world seemed to stretch before me with infinite possibility.

You wouldn't have thought I'd be a good candidate to fight fire, looking at me. I thought I saw my supervisor snicker as the crew boss came to ask him if I could be spared. I was a runner, pushing my body as fast as it would go on long lonely runs with only the harsh sound of my breath and the tyranny of a watch. I was unable to complete more than one pull-up. My arms were thin as matchsticks, my palms tender and soft. I still curled my hair. But the national park was desperate to fill out a twenty-person crew. I had taken the required fire school training, a few stultifying days parked in a classroom while look-alike boys in hooded crew sweatshirts droned on about relative humidity and something called the fire triangle. Everyone took the training, even fee booth workers at the park gates like me.

Some of my co-workers had gone out on fire assignments already, departing fresh-faced and returning with three weeks of sprouted beard. They swaggered around the compound, bringing fire stories to potlucks instead of food. They hunched over the picnic tables, desperate to escape their regular park jobs and go back out again. They were different people than when they left. I could see it on them. They had shrugged on confidence like a coat. They had been to war and back and it showed.

My supervisor said "You can take her," when pressed. "It's slowing down anyway." Then to me: "But you're laid off when you get back."

I was not at all sure I wanted to fight fire. There were very few women who did in the mid-eighties. I saw some of them at the warehouse, tough-looking types in ribbed white tank tops. Their hair was chopped off near their chins. They swung their faces aggressively over the pull-up bar. They walked big in the world. They seemed almost like men, no trace of softness about them. I admired them but I didn't see any way of becoming them. I was too slight, slender-armed like a sapling, someone who would break, not bend.

My secret was that I came from a lifetime of giving up: ballet lessons, abandoned when it was time to go on pointe, which seemed too hard. Gymnastics, tossed aside because of an overwhelming fear of falling from the parallel bars. Even riding a bike was a skill I had yet to master; I had given up when presented with a too-large bike absent training wheels, wobbling uncertainly for a few paces before deciding I'd rather just go inside and read. When things got tough, I didn't get going. "You just got so frustrated," my parents said by way of explanation whenever my non-bike riding came up in conversation, or my lack of skill at softball became apparent. Though my dad labored to build me my own gymnastics setup in the backyard, though he spent hours throwing a ball my way, both my parents saw how I turned momentary failure into a deep self-doubt. I wasn't able to shake it off like other kids. Each missed ball, every misstep on the balance beam became a reason that I wasn't good enough, that I wasn't like my schoolmates and never would be. Instead of trying, I retreated into the woods, a place that seemed to accept me for what I was, the forested backyard a calm and silent place far removed from fear.

I knew I was not the type of person who had a core of steel within, the kind that forced me to keep going, no matter what. I wanted to forge that steel.

I also wanted to fit in somewhere. In junior high I was an awkward girl, legs like pencils, not picked for teams, sent out to left field. The confident, sweatshirted girls in their clogs stalked weakness like panthers. They muttered loud enough for me to hear.

Beating her in the race will be a cinch.

Look at who's wearing high-waters! Expecting a flood?

Can't you catch the ball? What's wrong with you?

She's a wuss. I don't want her on my team.

The packs of mean girls drifted apart in high school, but I had learned by then that fitting in meant survival. But I never did fit in. I dreamed of mountains, not men. Rivers, not babies. I didn't know of any other girls with dreams like mine. When I left my hometown, most of the other girls had already anchored in for good, chasing crew-cut men from the air force base, their eyes firmly on the prize all of us were supposed to want.

Because I couldn't fit in, I ran.

I became obsessed with running, the kind of obsession where a fire licked my bones, a coal sat inside my skin. The slap of my sneakers on pavement was the only thing that mattered. Eight miles, ten, fifteen. I was fast, finally finding something I was good at doing, winning local races. Even then, it was never enough. I ran on the farm roads for hours. I ran through ice storms, tornado warnings, subzero temperatures, in the freezing, wet, cold, lonely dark.

I didn't know a lot about fighting fire, but I had learned a little from my fellow seasonals. I knew that to fight fire I would have to keep up with taciturn men who cut no slack. I would have to haul the same loads they did, stay up all night, hoist a combination axe and hoe called a pulaski for hours. I hesitated, torn. I had carefully avoided things that I thought I would fail at and I thought that I would certainly fail at this.

But this time I had a boyfriend. Jim was a dark-eyed motorcycle rider, the kind of man who appreciated daring. He was a local who worked a summer job at the park doing maintenance. This meant that he moved behind the scenes, fixing leaky roofs and picking up tourist garbage. When he drove through my fee booth, his truck piled with tools, he stopped and made small sweet talk until the waiting tourists honked their horns and his work partner rolled his eyes.

I didn't know if this was love or not. I had little experience with it; I had only flirted around the edges of it, clumsy and shy. In college I had chased after any punk-haired leather jacket would-be musician who would dole out pieces of his attention like secret jewels, then slowly fade as he sensed me falling for him. Sometimes it was I who would abandon them first, once they became needy, difficult to untangle from my life. In our senior year, my roommate planned her wedding and I looked on in equal parts longing and incomprehension. How could she know she would love him forever? What if some other man came along, someone better?

Jim was the elusive kind, the kind I always fell for. Out of my league, I thought. It was almost too good to be true, the touch of his rough fingers on my bare skin, the clutch of my heart as we accelerated around a turn, chased by logging trucks on the back roads. I was seized with the fear that it would all end, that he would see the real, flawed me past the exterior of cool that I tried to project. Or maybe it would go the other way: he would want to tie me down like my friends back home, comfortably married now years and years. What adventure could there be in marriage?

I knew only that I wanted to be different than the woman I was. I wanted Jim to believe I was a girl who would ride helmetless behind him on a motorcycle, a girl who flung herself headlong into life without ceaseless second-guessing, a girl who fought fire.

The person I imagined I would become by fighting fire was someone better: tan, long-braided, self-sufficient, strong. My confidence would blossom, confidence I had never experienced. I thought that if I became a firefighter, I could transform. I could jettison this soft skin, the name that didn't suit me, grow layers tough enough that nobody could break through. I would crawl inside someone else's skin.

I decided to go.

Before I went, the veterans gathered around to give advice. They tried to school me in the language of fire. They went through the steps of fighting a fire, ticking them off on calloused fingers: size up, scout, control, contain. They checked to see what I had packed in the red cloth bag that I would live out of for the next three weeks. Since you could end up anywhere in the country, you better be sure you have everything, they said. They advised head nets for Alaska, saying that the mosquitoes around the rivers were fierce. If I was going to the Southeast, they said, most times you motelled up instead of staying at a fire camp.

"Bring jeans!" one vet said. "You want a pair of talking jeans!" I would want them in town for the cowboy bars, to impress the boys if we were let out for R&R. "Rest and Recuperation, ya know," they said, "Mandatory days off after your twenty-one-day hitch, if they decide to keep you longer than that, if the fire's a gobbler. Sometimes you have to R&R at camp, and that really bites the big one."

I didn't know what a fire camp was. I was about to learn, they informed me, nudging each other and chuckling. "Earplugs, get you some of them. Generators, people snoring, hope you can sleep like the dead." There were actually many kinds of camps: the big fire enclaves, the small satellite camps perched somewhere far from civilization, called spike camps, and the coyote, where you laid your head on a rock and slept without your sleeping bag.

"Don't use a porta-potty parked next to a Sikorsky helicopter," they warned. "You might find yourself upside down when the ship takes off from the rotor wash. Those suckers have power." More laughter.

"Make sure you liberate all the stuff you can from camp when the Supply lady isn't looking. You know, lanterns, extra sleeping bags, tarps?" Supply ladies were stingy with their supplies, the old-timers warned, so I should bring a charmer with me to distract her. Some young kid with dimples would work. There was usually one of those on the crew.

They turned serious. "Don't smuggle booze into camp, that's a good way to be sent home." There was always someone who was blackballed, sometimes an entire crew. "Don't screw it up for the rest of us," they said.

If I was headed to California, the veterans said, I'd best not forget calamine lotion, because chances were I would get into the poison oak. "Nasty stuff!" they said, shuddering. Same for the Everglades, where I might be crawling through hardwood hammocks thick with poison ivy. A trip to the Salmon River breaks pretty much guaranteed an air show, where we would fly in helicopters instead of walk, so I'd need a chin strap for keeping my hard hat from flying away as I hiked up to the copter.

"Earplugs, for sure, don't forget," they said, ticking off other required items. Flight gloves, the sleek fire-repellent kind, so I looked cool in the helicopter, if someone would give them to me. Those were hard to come by. "Gotta be one or do one to get those gloves," they said, winking, speaking of the helitack guys who guarded the special gloves closely. On the other hand, if inversions set in, smoke lying low to the ground, I would be walking, so moleskin for blisters was a must.

"Take care of your feet," they warned. "Those aren't new boots, are they? You've got to break them in first. Step in a bucket of water and let them dry on your feet. If your feet get screwed up, you're toast."

"And for the love of God," they went on, "don't forget parachute cord." It could pretty much do anything: string up a Visqueen hooch in the rain, pinch-hit for shoelaces, stake out a recalcitrant tent, or tie up the hundred-foot length of slimy hose you had just carefully rolled. "P-cord rocks, go get you some."

They saw my blank stare. A hooch was an Alaska thing, if I got lucky enough to be sent there. In the interior, crews took rolls of black, shiny Visqueen, sliced off sheets with their knives, and rigged up a low-rent tarp in case of rain. Alaska was the best, if you could swing it onto a smaller fire, off on your own with a little crew, a shotgun for bears, and no overhead team on your back. Their eyes grew hazy with decades of memories.

The veterans' advice was sprinkled with words I didn't understand. They talked about avoiding slurry drops, those red-tinged cloudbursts that came from the belly of an air tanker to stop the fire's advance. "Drop your tool and get the hell out," they said. "A big drop can kill you." There were also cat-faces, trees with their bases burned out, that could fall on you while you ate lunch underneath. If someone yelled "Bump up," that meant you were taking too much time with your patch of ground and you needed to move along. If someone yelled RTO, that meant to Reverse your Tool Order and run to a safety zone, packs and saws be damned. There was a whole section of dialect that went with hose lays: spanner wrench, gated wye, forester nozzle.

The same with helicopters. Only rookies called them choppers. For some inexplicable reason, helicopters were called ships, and there were many different kinds, from tiny reconnaissance to the gargantuan ones that lowered a snorkel into the water to fill up. Others hauled buckets almost as tall as I was. Either way, you stayed out of their rotor wash.

I had learned about the various stable of tools in fire school. I knew about pulaskis and shovels and the rake-like McLeod, but now the veterans told me there were others I might use. Rhinos were hoes that could be used to chop stubborn soil. In Florida, people used flappers, which looked like the name suggested: long-handled with a rubber flap on the end for smothering grass fires. Some grassland fires even were fought with wet burlap sacks. You picked the best tool there was and just went for it.

Hotline, where you paced right alongside the flames, digging as fast as you could, was a rush unlike any other; mop-up, where you drowned remaining smokes with water and scraped out embers far into the fire interior, was its own kind of boring, sweaty hell. Being assigned to rehab—breaking down dozer piles, reseeding burnt areas, raking firelines—meant no overtime or hazard pay, both additions that could fatten a paycheck. Burning out—setting a backfire to stop the oncoming flames—meant you got to drag a drip torch, a squat metal canister filled with unleaded gasoline and diesel. You lit the wick and set a fire that would cheat the main fire of fuel. That's the best job ever, they said. Other jobs you might get included cold-trailing, running an ungloved hand over burnt ground to feel for heat, or running a pump. You might have to hike a long ways to fill up a piss pump, a bladder-like water container with a spray nozzle, and hump it back up to the fire. "Dry mopping," they reminisced with a shudder, recalling those remote fires with no water source, where they had to stir soil into the embers until the heat was smothered.

"Keep one foot in the black," the veterans concluded, meaning places that had burned already. They unloaded their gear from the bus that would carry me to my fire assignment. They were unshaven and dirty. Their eyes were sunken and red-rimmed, but they watched me with envy. I could tell they couldn't wait to get back out again. "Two days R&R and we'll be back!" they called. "Save some fire for us!"

"Living the dream," someone on the bus said. And it did seem like a dream, a steep dizzying dive into murky water. I had heard from the veterans that you either loved fire or hated it—that it could get into your blood and you couldn't shake it. You were doomed to follow fire across the country, east to west, north to south, they said, grinning. If you hated it, of course, you could stay safely in the visitor center or the fee booths, clean and warm but minus the extra pay. Those posy-pickers couldn't cut it, the veterans implied. They weren't tough enough. Not like us. I wondered which one I would be.

We drove late into the evening, small towns glowing far out along the horizon. Our orders were to sleep, because we would be out on the line in the morning. There was no way to know where we were going; the crew boss snored in the front seat, his big feet bare. The others passed rumors back and forth along with dips of Copenhagen. We were going to Boise, Idaho, to the interagency fire center, Doug, the crew know-it-all, whispered. We would stage, sleeping on the lawn, until an order came through for us. Someone else had heard that we were headed to the Silver Fire, a big gobbler straddling the border between Oregon and California. The last crew had just come from there with tales of fire in the treetops. The veterans spun fire stories from faraway lands: Alaska, Kentucky. Each one was more dangerous than the last. They spoke in low voices, words forming on their lips like pearls. I leaned in closely, trying to absorb what they knew, what they had seen.

It was impossible to sleep on a shockless school bus, already smelling of us, our brand-new Nomex, our half-eaten tuna sandwiches. Some of us clicked on our headlamps and read. Others played cards across the backs of the seats. The lucky few dozed, their legs hanging out in the aisles, their bodies sprawled in impossible contortions.

True to Doug's prediction, we first spent a night sleeping on the pavement near the Boise interagency fire center, rows and rows of us like a line of baked potatoes in flimsy yellow synthetic sleeping bags, hundreds of people staging, waiting for an assignment. Sleepless, I roamed the darkened sidewalks. Others walked too, a flame in the night from a lit cigarette, a deep cough from the shadows. We nodded as we passed each other. I wanted to stop some of those roaming strangers and ask them something. Anything.

What's it like out there?

Are you afraid?

Can someone like me really do this? What if I let my crew down?

In the end I was too shy and mumbled something as I passed, finding my snoring crew where I had left them.

The next morning Doug pointed out the THANK YOU FIREFIGHTER signs, hand-written and posted crookedly on the side of the road. "Hey, that's us!" he said. People in passing cars honked and waved at us. We waved back, smiling broadly. It was impossible not to feel like I was part of something big and brave, a foot soldier in a glorious war.

On the fire where I ended up, though, doubling back to the Blue Mountains of northeast Oregon, we were far from any houses or news channels. It was called the Ryder Creek fire, named as usual after a nearby geographic feature. "You're kidding me," one of the vets moaned as we pulled into fire camp. "Nine thousand lousy acres? Look at this cluster." He was one who had regaled us with tales of massive fires the night before. Over sixteen hundred fires were burning in the Northwest, he babbled, and they were calling it a hundred-year event. We wouldn't see this again in our lifetimes, he rhapsodized. These were stand-replacing events, huge fires that destroyed whole forests.

The camp seemed like an afterthought, set up in haste in a stubbly field. Army tents flapped in the stale breeze, a crew to each tent, straw spread over the ground in an attempt to even out the lumps. Crews had drawn their names on cardboard signs: SHO-BAN #2, KENTUCKY REGULARS. Most of them were out on the line, their red bags neatly packed in the shelters. You always had to be ready to go somewhere else, Doug said. You didn't have your shit scattered all over creation. The crew wouldn't wait, and you would be left behind at camp.

The bus had pulled up to a row of other buses like ours. Alec, our crew boss, motioned to us to stay onboard, although a brave few made a break for the line of porta-potties. He disappeared under a tarp but soon reemerged, shaking his head and holding a hand-drawn map. "Gear up and go," he reported. "Drop Point Ten," he told the driver.

Drop points, I soon figured out, were wide spots in the road where crews were let out and picked up again. We passed a couple, marked with numbers on cardboard. We rode with our faces pressed like moons against the window, eyes wide at the country dropping off abruptly near the tires. This was wrinkled country, ridges fanning out like crow's feet around a man's eyes, lodgepole pines marching up the slopes as far as we could see through the unwashed windows. Fearless, the driver took the turns fast, the dirt road barely wide enough for another rig to pass. Every campground we passed was closed, the tourists booted out. Top-heavy engines trundled down the road, their occupants yellow-shirted blurs, the mirrors of our rigs nearly touching.

The bus wheezed to a stop at a clearing and we stepped out at our drop point, blinking in the flat sun. A dozer sat abandoned, but otherwise we were the only people here.

"Gear up," Alec said. He changed out of his sandals to tall fire boots. "We've got a granola for a crew boss," Doug whispered. I didn't really know what that meant, if that was good or bad. Good, I decided, used to being labeled the same because of my fondness for nature and Velcro-strapped footwear.

Alec spit on the ground; we were too slow, too inexperienced. There were twenty of us, the normal handcrew number, three women, seventeen men. A crew hastily cobbled together from the park ranks, we had only been to a handful of fires between all of us. Fully a quarter of us were rookies, on our first fire.

We were assigned to three squads of five people each, with a boss heading each squad up. Alec ran herd on all the squads, with a trainee under him, learning the ropes. My squad boss was a skinny guy with deep wrinkles across his forehead. He appeared deeply uninterested in us, huddling with the experienced guys instead, casting a withering glance at the rookies. It was obvious he felt he had gotten the short end of the stick, three rookies on his squad.

We grabbed for our tools in an unorganized knot of nervous energy. "Hustle!" Alex bellowed. I hastily pulled out a pulaski, a tool I had only handled once in fire school, stepping on the toes of one of the guys, who groaned to see only a shovel left. He tried to trade it, but those of us with pulaskis hugged them close. I had heard that if you got stuck with a shovel, you might as well pack it in. You would be parked at the back of the line clearing out massive piles of dirt left by the rest of us. I wasn't sure of much, but I was sure I didn't want a shovel.

Tools collected, the crew assembled in a messy clump, waiting direction. A few people tipped back canteens or pulled out granola bars. Others fussed with their packs. These were all things that could have been done on the bus or back at camp. Alec sighed, combing his tobacco-stained fingers through his long beard. We were supposed to line up as soon as we had tooled up. We were supposed to look ready, not disorganized. He had told us this before. There was an order to everything, and we had failed. I could see it in his disappointed expression. He didn't say anything, but I imagined that he was thinking that it would be a long three weeks.

Doug forgot and put his tool on his shoulder, a big no-no, and got chewed out. Only the sawyers were allowed to do that, the saw wrapped in a pair of chaps to prevent the teeth from chewing up their shoulders. If the rest of us did it with our unsheathed tools, it could take out somebody's eye, or so Alec told us. Doug winced under the criticism. There were no manuals for this type of work, only unwritten rules that you learned as you went along. Being yelled at was part of learning, it seemed.

"Line out," Alec hollered. We fumbled at this too, pushing into a ragged semblance of a straight line. I wedged myself behind Doug and in front of a curly-haired guy whose name I didn't know. At the front, Alec set a brisk pace.

The forest was dry, twigs snapping under our boots. Fallen pine needles coated the ground, sending up scented dust and the dry crackle of drought with every step. We headed straight uphill through a crowded forest, the slim trunks of lodgepole trees blocking out the sun. There was no trail. Alec stopped often, consulting his map. He scratched his beard and pulled out a compass.

We hiked single file, heads down, talking little. Only Alec knew where we were going, or at least I hoped he did. We had no idea where the fire was; he kept that to himself. I didn't see or smell any smoke. This was the way things worked, Doug whispered. We weren't meant to know anything. We just followed.

"With crew bosses on fires, it's his way or the highway," he hissed. "And if you don't keep up on the hikes you will be sent home."

What if I couldn't keep up? Already I was sliding on the slippery needles. My socks bunched under my heels. My unfamiliar boots pinched at the toes and I could feel blisters beginning to blossom. All of the childhood failures loomed closer; it would be easy to give up the way I always had. *No,* I thought, *not this time.* Instead I scrambled in a half run, heart pounding, my pulaski in a death grip.

Hiking to the line, we sometimes climbed over deadfall higher than my head. "Jackpots," Doug said, slipping over with ease as I struggled to balance from trunk to trunk. These timber downfalls were firefighter killers, he said, places that you didn't want to be if they went up in flames.

I didn't want to be anywhere on this parched hillside. The trees closed in on me, their branches snagging on my clothes and pack. The only sounds were twenty pairs of boots pushing through mounds of dry pine needles, banked like snowdrifts beneath our feet. Even my untrained eyes could see that this hillside was dangerous, the skinny trees a snarl of dead branches from top to bottom. There was a fire in here somewhere and we were walking toward it.

The guys outpaced me and I raced to catch up, my breath ragged and hot in my throat. I did not want to be alone, not here.

At one point Alec discovered we had taken a wrong turn and would have to backtrack. He puzzled over his map, drops of sweat pockmarking the wrinkled paper. "Reverse tool order!" he exclaimed heartily, meaning that we had to turn around and go back the way we had come. Because of the mistake, we would hike six miles before we even got to the line. Alec shrugged. "Pays the same," he said, changing course.

We headed up in another direction, but I was not sure how he knew the difference from where we had been before.

I was used to trails carving through a landscape, trails people had built. This type of navigation, using a compass and a squint into the sun, was new to me. We churned through the forest for a while longer. Scratchy transmissions blared from Alec's radio, bobbing on his belt. I caught snatches of conversation—"Copy that. I'll tie in with you on the ridge." Overhead, a helicopter droned unseen.

Alec finally stopped and the rest of us straggled up behind him. I fumbled with a canteen and gulped down warm water that tasted of plastic. We were standing on a saddle, our boots hanging over the edge of a low, sway-backed point on a tree-lined ridge overlooking lower ridges that undulated like waves toward the horizon. All I could see was forest, an unbroken stretch of shades of green. I had never been this far from pavement before.

"Okay!" Alec boomed. "Here's where we tell the men from the boys." He sized us up, our chests heaving as we struggled to breathe, our hard hats askew. Like a drill sergeant, he paced along our line until he found a fire veteran. "Tim, you're the lookout."

Tim, slight of build and bushy of beard, nodded and set down his pack. I remembered from class that we were supposed to have eyes on us while we were on the line. He would watch the fire below us because we would not be able to see it from where we worked. He would warn us if there was a blow-up down below, threatening us and our fireline. Armed with a radio, he perched on a rock outcrop while the rest of us dropped into the forest.

Our fireline would go just below the saddle, a place that Doug, who appeared to have adopted me, said would be a good spot to stop an uphill fire. He pointed out the way the trees grew sparser here, making less of a meal for the fire. Smoke filled the bowl below us with a hazy blue. We were following a flagged line someone had scouted. There were no other crews in sight. We were all alone. There was no fire, just the blanket of smoke. Around me everyone was unsheathing their tools as if this were normal, to dig a line in a random section of forest.

"Let's go," Alec said. "We're going indirect. Away from the fire," he explained to my puzzled face. "The fire's down there, in the draw, but we can't dig line down there. Too dangerous. The big kahunas want a line here." He waved a hand at the steep bowl below. "Got two miles of line before we tie in with the hotshots. Let's get a move on. I just want to see asses and elbows!"

Catching my breath, I covertly watched the others. On the last day of fire school we trooped out to a field and practiced digging fireline, but it was loamy dirt on flat terrain, nothing like this. I didn't want to screw up, so I watched while slowly taking the plastic sheath off my tool.

Leaning over, the crew swiped at the ground, pulling the loosened material away to one side and mounding it up. This would be the "green" side, the one the fire was not supposed to cross. Like slices of cake, chunks of pine-encrusted soil flew each time someone brought a tool down. The developing line widened to about three feet, a slash in the heart of the forest.

The guys who knew how to do it used their bodies to their advantage. Doug rested his digging arm on his bent knee, using the power of his muscles to move earth but also to rest as he went. This was better than some of us rookies, who flailed at the ground, arms windmilling, legs churning. "You're taking too much!" Doug screamed at a rookie ahead of us who was laboring too long on one piece of ground. The rest of the crew had bunched up behind him, waiting. "One lick and go!"

I grasped my pulaski in sweaty gloved hands. As I swung it high over my head and brought it down, the hoe end struck a rubbery root and the tool bounced harmlessly into the dirt. Uneasily I looked around, my cheeks flushing hot. But nobody had noticed. Switching to the axe end, I hacked at the root until it broke apart.

"That tool's paid for," Alec said as he bustled past, meaning that I was clutching it too closely to its end instead of farther back on the handle. Embarrassed, I changed my grip and moved further along the line.

We worked mostly in silence, each of us in a bubble of our own thoughts. Though we labored only a few feet apart, I was aware only of blurred motion beside me, the chink of someone's tool striking unforgiving soil. Disjointed scraps of song floated through my head. Ahead of us the chainsaws whined in uneven harmony. Two swampers, guys who cleaned up after the sawyers, scurried back and forth between the sawyers throwing limbs off the line. Then there were the rest of us, bending and swaying in a strange kind of dance.

The next day we would do it all over again, and the next, until twenty-one days were done, or the fire was out, whichever came first.

The fire moved closer as we worked. I brushed sweaty bangs out of my eyes, trying to see. My bandanna did little to filter the thick smoke that had suddenly enveloped us in a gritty gray cloud; I could feel it settling deep in my lungs.

Somewhere unseen, the fire moved below us like a living creature. Occasionally I could hear it crackling through the dry trees far below us, but it was blocked by the slope. The crew line snaked around the corner and down the hill, out of sight. This far back, I could barely hear the growl of the chainsaws. There was only my breath in my ears, a ragged pulse of sound. The chink of my pulaski as it met rock. The grunts of the others who bracketed me as they worked.

"The fire's just below us! Past the trigger point!" Tim called on our secret radio channel, the one all the crews had so we could talk privately. His voice was tinged with panic. Alec ran by faster than I thought such a burly man could move. He had been babysitting us, standing over us with a shovel while we worked, not altogether pleased with our performance so far. Roots meandered across our line, paths fire could follow into the green. We had left limbs on straggly trees that could become firebrands. It was the luck of the draw what kind of crew you got, and he had drawn us, a motley combination of trail crew workers, naturalists, and rangers. Now our lookout had disappointed him too by not paying closer attention to what the fire was doing and notifying Alec earlier. Standing helpless with my pulaski, I grasped that the fire was coming uphill toward us and we needed to get out now.

"Back to the drop point," Alec ordered, setting the pace.

I paused for a second before we bailed off the ridge into the unknown, abandoning our line. I still could not see the fire from there, but I could hear it. It sounded like the howl of wind through a forest, a wild storm tossing branches and toppling trees. The smoke thickened to a brown soup.

The guy behind me stepped on my heels and I hurried to close the gap. We did not run, but double-timed it. A current of fear surged down the line. I was unsure whether we would burn up any minute or if we had a few moments of safety. Trying to keep up with Alec, we vaulted over logs and scrabbled across downed trees. My ankle buckled but I was up and walking fast. *Keep up, keep up,* I whispered to myself, a mantra. I was terrified of being left behind.

Running on a fireline was an admission of desperation. Even the rookies knew this. Heat from the flames below preheated the slope ahead; fire could skip over the terrain, gobbling up timber at acres per minute. What we could not see could kill us too: superheated gases in advance of the flame front could literally suffocate us. It was foolish to try to outrun a fire. Although we knew this, had learned it in fire school, we picked up the pace until we were nearly jogging. Nobody dared to look back.

Alec led us directly down the mountain. It took less than half the time it did to go up. In the short time since we had come this way a makeshift trail had formed, an eroded dusty goat path that gave no concession to the terrain. Instead it plunged down the mountain without mercy, delivering us to the drop point in the minimum amount of time. No such things as switchbacks here. Our goal was to get up and down to the fireline as fast as we could.

We broke out at the drop point on the dirt road, sun flooding across the cleared dozer line. I stood there in confusion, not sure what to think. Were we heroes or cowards? I didn't know. I looked to Alec for clues.

"Blow-up," Alec said. "Fire's taken off in the heat of the day. Typical." He spit.

I looked up the way we had come. Fire was up there, probably crossing our line, but I could see nothing but smoke. It rolled over our drop point, choking us as we put away our tools. From here it seemed harmless, but as I watched the smoke parted and a column pierced the sky. It was massive, an anvil shape alien to the otherwise oblivious blue canvas above us. It was both terrifying and somehow exhilarating. I wanted to be back up there, I realized, in the heart of the fire.

I could see that the others did too. Their faces were upturned, teeth a white blaze in blackened faces. This was what would hold us together in the twenty-one days that we would be on the line as a crew. We were united in a way that was not possible outside of the fireline.

The bus driver roused himself from sleep, cranking open the door. We filed on, already grinning, making up our first fire story. In camp, we paraded around the mess hall, flaunting our dirty faces. We felt baptized by fire. "We were in the blow-up," some of us said to whoever would listen—the servers in the chow line, the lady at Supply. They observed us nonchalantly. They had seen all this before.

Alec sighed again, watching us. We had so much to learn.

Nobody mentioned that we could have died on the ridge, like other crews had in the not so distant past. Trapped by a fire we couldn't see coming until too late, we could have baked to death even in our flimsy fire shelters, the foil and aluminum tents we all carried. If any of the rest of the crew thought this, they quickly discarded it. I did too. That happened to other people, not us. Alec would keep us safe.

The overhead people down in camp knew what they were doing. We were just a Type II crew, a call-when-needed crew. Bad things didn't happen to those kinds of crews. You had to majorly screw up to get trapped.

Our line abandoned, we were sent to other ridges and other firelines. One day we stopped at the helispot, a small opening that had been cleared of trees and brush so that the helicopter could land with the gear and tools we would need for the next three weeks. A big pile of rolled hose, cardboard boxes of Meals, Ready to Eat and portable pumps were piled in a messy bundle on top of a cargo net. We quickly took out what we wanted for the shift. The guys grabbed the dolmars—red plastic containers with two compartments, one with saw gas, one with oil—and extra falling wedges for the sawyers. Doug took a handful of files and shoved them in his pack.

I looked at what was left: a cubitainer—cubie—of water. It held five gallons, and I did the math in my head: forty pounds. A plastic jug encased in cardboard with a carrying handle. The only way to move it was to hold it in one hand.

It was a scorcher, with temperatures expected to soar into the eighties. We were fighting fire in a parched lodgepole forest. There were no creeks nearby and we would be digging fire line for at least sixteen hours. We would be thirsty, having sucked down the four quarts we carried up here. We needed this water.

The guys had already started up the slope. "Close the gap!" Alec yelled. I looked around but there was nobody else to carry the water. As I hesitated, a grimy crew appeared. Their matching hard hats labeled them a hotshot crew, a

Type I, the kind that worked together and trained hard all summer, not a pickup crew like ours. In the fire world, they outranked us.

There was one woman with them. She was tiny, barely scraping five feet, her long dark hair in a braid. She smiled as our eyes met, mutual recognition of the firefighting sisterhood. I noticed that she was wearing her pack and had two full water pump bags slung over her shoulders. Besides her pack, she was carrying eighty pounds of water.

As her eyes flickered over me, I thought that she could surely see right through me to the scared girl that hid beneath. What else did she see? Did she see my soft core, the worry that dogged my every step, the fear of failure that snaked through me, the desperate desire to find a place to belong? Or maybe she only saw this: another woman in a world ruled by men, here for some reason that most people would never understand.

I watched as she walked easily down the slope in the opposite direction of my crew, her slight figure bowed a little under the weight. I saw so few women out here that I wanted to chase after her and ask her how long she had been out here. How she kept climbing the steep hills with men twice her size, how she kept swinging a pulaski for hours on end. Instead I looked back down at the cubie. The familiar emotions of frustration and despair pushed their way in. A treacherous part of me wanted to give up. Voices from my past surrounded me.

This is too hard. I can't do it.

I'm not good enough, not strong enough, not brave enough.

The hotshot woman had disappeared into the trees. Who did I want to be? The same girl I had been? Or someone new, someone who walked through the world without looking over her shoulder? I drowned out the voices and picked up the cubie.

The walk up to the staging point where we would start to dig line seemed like miles. The plastic handle dug into my gloved hand, water sloshing around in the container. I bent over when nobody was looking, covering ground in a crab-like run that stabbed at my knees. My arms ached. The guys were far ahead, the blaze of their yellow shirts only occasionally catching the filtered sun. Finally we reached a line of double flagging, showing us where to begin. The fireline would stretch downhill into yet another brushy canyon, red flags dotting the lodgepole pines far below us.

I set the cubie down with relief. The handle had bitten deep into my skin. I would have to bump it down the line with me until it was empty, because nobody wanted to waste energy climbing back up for water. I would have to balance on a steep slope as I descended, pulaski in one hand, sloshing cubie in the other. I wanted to leave it for someone else. Then I thought of the woman with the piss pumps, somewhere down below us, and I knew I could do it.

Time had no meaning on the fireline, but somewhere in the afternoon we broke for lunch. It was possible to do this only because the fire was quiet today, chuckling to itself as it slowly crept through the piles of fallen trees below. Slowly the season was changing in these mountains. The days started out with frost etched onto the meadow. The fire was weakening against winter's advance.

Our shift stretched into the evening, and we pulled out our headlamps. A few stars pricked the growing darkness. Trees, weakened by fire or disease, fell somewhere near us, unseen and terrifying thunderclaps in the forest. We had not reached the main fire yet today, though I thought I could see a faint orange glow somewhere below us. As usual, our line did not look beefy enough to stop it.

Alec hustled by from some unknown place. "Huddle up, we're heading back to the drop point as soon as the night shift gets here."

We gathered up together, sucking down the dregs of our water. We were spent, a formless mass of bodies pushed to their limits.

The night shift hiked up the hill, looking fresh. They showed no signs of sleep deprivation, although it was a sure bet that the generators and school buses had kept them in an uneasy state between sleep and waking. They were mostly men, a Shoshone–Bannock crew. They nodded as they approached.

"Welcome to hell," Doug said, his boots off and his feet plastered with moleskin. The Sho-Bans only nodded. This wasn't their first time on the line.

We geared up, gathering crumpled lunch bags and water bottles and tools. In single file, we headed back down the ridge. The next day we climbed back up, encountering the bleary eyes of the night shift. By default I still hauled a cubie, learning that if you had it once, you might be stuck with it for the entire fire. Digging line day after day, we were lulled into complacency, our world shrinking down to the square of earth in front of us. It seemed like we would dig like this forever. I had come to a place where the blisters on my hands no longer hurt and I could ignore the ache in my back. At the same time there was something deeply satisfying about the unclenching of my fingers each night after setting down my pulaski, the strike of metal on metal as I filed its blade to a sharp edge, the curl of silver shavings on my pants. I sat in a long line of firefighters bent over their tools, the sweat chilling our backs as it dried in the setting sun. This was my life now. Everything else—Jim, the park, the place I had come from—had been left far behind. I was seeing myself in a different way, as an athlete, a link in a chain, far from the girl I used to be.

Every day melded into the one before. Alec rousted us with a hoarse shout. Grimy faces poked out from sleeping bags. It was four thirty, barely light. We choked down lukewarm scrambled eggs in the mess tent and stumbled outside into a world the color of steel wool.

The siege went on into October. At night the temperature dropped to single digits. My contact lenses froze inside their case. The wall tent we slept in did little to shield us from the cold. Alec found us paper sleeping bags to put over our slippery fire-issue ones, so we wriggled inside several layers of cotton and synthetics, contorting like snakes to get comfortable. Still, it was impossible to be warm.

I changed inside my bag in the half-light, but the guys didn't care: they stood shivering in their droopy boxers, pulling on their pants. We all looked the same, androgynous with our matted hair and shapeless clothes. We started our hikes in long underwear and sweatshirts, shedding layers as we walked and the day heated up.

The days were punctuated only by the breaks we took to gulp down water and a sack lunch provided by the caterers, or enforced stops to replace a chain on the saw, the crew backing up behind the sawyer and swamper. The hours passed by in a seamless rhythm.

Occasionally someone roused himself to start a group discussion.

"What was your favorite toy when you were a kid?"

"What's a word that rhymes with orange?"

"Who else has monkey butt from these pants?"

Mostly we just talked about fire.

"What did you burn down as a kid?"

A hay field, a barn: the answers came quickly from some of the guys. When I stared at them in disbelief, they shrugged it off. "Guess I'm just a pyromaniac. Aren't all firefighters?" It was hard to know if they were kidding. The veterans liked to tease us rookies, whipping us up into states of confusion or outrage with gritty stories: the questionable women they had dated, the adult shows they had frequented, the bugs they had eaten on the line for entertainment.

Sometimes we dug right next to the fire, heat flushing our faces like sunburn. We dug without talking, urgency lending a second wind to our swings. Other times we mopped up a sullen, dirty burn, dragging kinked, wet hose over logs and around stumps to extinguish stubborn smokes.

We never knew where we would go from one day to the next. Alec appeared after briefing, his sweatshirt hood tight around his face. The only thing that showed was his beard and his eyes. Waving a shift plan, he told us to line out and we did. We didn't ask questions. It was hard to say if we were gaining on the fire or not.

At the showers I saw myself in one of the mirrors above the sinks. I was unrecognizable, someone with eyes that blazed and a new set to her jaw. I looked for a while at this stranger until Doug nudged me. "I just took a whore bath," he informed me, waving a dirty wet wipe. "I'm not taking a shower until we are done with this fire."

"Gross." Like the others, Doug often tried to shock me with trash talk and lack of hygiene. Some of the others claimed to wear the same pair of tighty-whities for the entire assignment. The trick, I had learned, was to act like it didn't bother me. Alec occasionally roused himself to shout that there were ladies present, but that only made the guys worse later on when he wasn't around.

Because there were so few women in fire camp, we didn't have lines at the showers like the men did and could go right in. Black soot poured off our bodies as we showered, stomping our gray underwear underfoot in a futile effort to clean it. We lobbed shampoo and razors over the stalls to each other. My hair froze in a frosty halo around my head as I scurried back to the tent.

Finally, incomprehensibly, we were done. Instead of palming the cold metal of a pulaski and hiking up a mountain, we were sent home. The fire still burned, but the skeleton crew left behind could handle it.

"See you on the big one," Alec said outside the demobilization tent, where we went to sign off on our timesheets and arrange transportation home. It was clear that he never expected to see any of us again. Although we had not screwed up too much, he hoped for a more hardened crew the next go-around. He saw so many rookies go out on one fire and then quit. I suspected he thought that the same thing would happen to us.

Alec whistled as the bus took us away. He was staying to work on the rehab crew. Even though he would be breaking apart berms and running a seeding machine, tedious work without glory, he was glad to be rid of us.

At the park, the fee booth was shuttered for the season. The winter rains had begun, a low ceiling of thick gray clouds clamped firmly over the town that would not lift until summer. The crew disbanded at the warehouse without any backward glances. They got into rusting pickups and sped off. Suddenly our bond was gone as if it had never been. I stood for a moment with my red bag, bereft, the drizzle needling its way into my skin.

Jim avoided my eyes as he talked about college. He had decided to try it again, hoping to avoid a career of picking up trash at the park. "I'm too young to settle down," he said. I wasn't sure that I wanted to settle down either, but I hadn't expected him to strike out on his own so soon. I had expected a few more seasons out of him, more motorcycle rides, maybe love. His rejection stung. He didn't see the new me, the woman who had marched up and down hills with a pulaski. He didn't see her at all.

There was an ache in my heart worn deep over the next few weeks of watching him slowly leave. I thought about my friends back home, chained to their lives. They seemed content with their choices, but their eyes followed the coal-laden freighters across the straits. They had traveler's souls, but not the desire to propel themselves out into the world.

It was time to choose. I looked at my road atlas. It was full of possibility. I had heard that you could patchwork a life out of seasonal work, making long zigzags across the country. It could take you a year to work your way back to your starting point.

The seasonal workers I had been on the fire line with packed up their trucks in a cold rain. They crawled underneath like monkeys, changing oil. They jettisoned belongings. *Want this bike? This set of free weights?* They were animated by the thought of moving on. They were going to Texas, the Keys, Colorado. They might be back here next year, but then again, that was six months into the future, impossible to predict so far away. Beers in hand, they skillfully placed their entire lives into their vehicles, carving out a small place to sit.

They told me that it was possible to drift around the country from place to place. You just got a seasonal job in summer and saved up as much as you could to last you through the winter. You could get a job at a ski resort loading butts onto lifts if you had to, or you could travel on the cheap. Peru. Mexico. You could work both summer and winter for the parks too. There were even people who cobbled together winter and summer seasons by going to Florida to work on a fire crew for a few months. Fire year round!

The best thing about it was, they said, that if you don't like one job, you only have to stay there for a few months. Then you can move on to something better. You aren't locked in. To anything!

You could really do this with your life, I realized: travel with a colorful group of people around the country in a long unchanging dream. You did not have to grow up, get a mortgage, get married, labor away at a job you grudgingly tolerated just so you could save for retirement. Retirement was years away, decades. Instead you could live in a national park, sleep beneath the sequoias, wake up to the cloudy peaks outside your window every morning, hike up to where the rivers begin their life high in the same mountains. Who wouldn't want that?

A thought kept running through my head: During every month in the country, a fire was burning.

"The National Park in Nevada is hiring," Doug said. He was headed there himself in a low-rider car with a bad muffler. The jobs consisted of leading tourists through a cave, armed with puke bucket and heavy-duty flashlight, but he was sure that I could weasel my way onto a fire or two if I wanted it badly enough. I hesitated, thinking about my student loans. I barely got paid enough to cover the meager rent on our bunkhouse, much less make a dent in the debt that hovered over my head.

Doug just kept packing. He shrugged off responsibility as easily as he shucked off his Grateful Dead T-shirt for a fresh one. This was the life, he implied; we were young, why wouldn't I go?

I thought about the hotshot woman I had seen this summer. Where was she? I was sure that she wasn't holed up at a full-time job in a fly-over state. She wasn't pining after a black-eyed motorcycle rider who had left her. Nobody pinned her down, and she had no regrets. "I'm putting in an application," I told Doug, rushing into the bunkhouse to find the government form. The call came only a couple of weeks later, after most everyone else had left and I was volunteering my time at the visitor center.

I slid into my Chevette and sat there for a minute. I was alone with a mixture of fear and regret and desire, all mixed up together in an emotional stew. It made me both punch the gas and tap the brake. Jim was long gone, his heart closed to me. I had never felt as alone as I did right then.

It's always like this on your first drive, the veterans who were still left said. It will get better. They hit the hood with light fists. Drive on, sister, they said. We'll see you on down the road.

Sisterhood

On some nameless dry mountain in Wyoming, a girl cried. The first day of hiking to the fireline, hours uphill, did her in; she collapsed in a slender heap. "I'm exhausted," she moaned.

All day long my crew had chased Dave, our sawyer, up the mountain. Dave carried the big saw on his shoulder as he hiked, smirking at the rest of us when we couldn't keep up. Like terriers, we dashed after him, but he outpaced us with his long legs and determination. When we finally caught him, he was calmly sitting under a pine, smoking. "Steady goin'," he greeted us with a flourish of his cigarette.

We took a breather, the crew scavenging the sack lunches for chocolate and tossing the rancid ham sandwiches. "Pigs are gonna die," Dave hooted. This was his prediction that this fire was going to last a long time, thus weeks more of ham sandwiches.

Starving, I joined in the lunch bag raid. "Don't high-grade the lunches!" our crew boss yelled, but I ignored him. We hoarded food, all of us, stashing it in pockets and pouches. We lived in fear of running out of food. I knew this because our conversations were all about food—what we would eat when we were done with this fire, our favorite meals, what wasn't in our lunches but should be. Our bodies burned calories with a fierce fire. Paydays and Skittles arced through the air, long-armed men snagging them in one graceful grab.

I no longer feared food the way I had in college, the way all of the other girls had, too. Food then had been our enemy, divided into categories of good and evil. Whittling my body down to an arbitrary desired weight had required constant vigilance. I woke up in the night regretting what I had eaten, cataloging every bite, believing that I had somehow expanded, ballooning wide, fat seeping through my blood, bulging out my belly, my thighs, my butt. I had to get up to check, turn sideways in the mirror for reassurance.

Fighting fire ignited a different kind of burn inside of my body. My hunger said yes to the sugary sweetness of Oreos snatched from their plastic sleeve, yes to the creamy filling from the chocolate wafer. Yes to the sandwiches two at a time, yes to the pillowy white bread sticking to the roof of my mouth. Yes to the brick-like granola bars I pocketed, caching them as insurance against future hunger. My body unfurled like a plant long denied the sun. I grew an inch in height over one summer.

It was only my fourth fire, but I already knew that nobody cried. Any weakness you had was shoved down deep, replaced with a hard shell. People who cried were the same people who would give up halfway through a shift. They would leave rubbery roots in the soil that gave the fire a foothold. They would slow down a double-time march to safety. They could burn up. They could die. All of us could die because of them.

What you did instead of crying was swallow the bitter taste down deep where nobody could see. I was familiar with the hot prickle of unshed tears, but each time a crew member poked fun at my ineffectual pulaski swing or challenged me to hike faster, a new, stubborn determination rose in me. I could almost feel it, a thick slurry filling up all the weak spots so that all that remained was pure, strong granite.

When we started back up the mountain toward our spike, the girl lagged far behind. The crew boss looked grim as he stopped to wait. We all knew it. She was the weak link, one that could endanger all of us.

As though she were an injured animal, the other women shunned her. We turned our backs, gathering together in our spike camp, peeling bloody moleskin off our heels and upending cubies over our shampoo-coated heads. There was a feeling of danger, as though the men would turn on us for the mere fact of our being women. We avoided looking at each other. Each of us knew that one woman crying could undo each night shift we endured, our arms boneless as rubber, each climb spent running to keep up, everything we had done to earn our place here. We did not want to be associated with weakness of any kind.

It felt both cruel and necessary. I had heard the under-the-breath muttering from the old-timers, the ones who fought fire for years before women came along. The jokes about how we must be on the rag if we acted surly, the deliberate forced marches so that we had to struggle to keep up, the blatant trash talk—those were all familiar to us. The men with us now, some of them, were old enough to remember when women were actively discouraged from fighting fire. Some of them still didn't want us here.

The girl was sent back down to the main camp, where she joined a helitack crew, loading up helicopters with gear for the spike camps. Not real firefighting, the men implied. A good job for a woman. She wasn't tough enough, we told each other, putting on our game faces. I believed that we all wondered what would make us break, but I was glad she was gone. It was hard enough, an obstinate fire that defied our lines day after day. The men eyed us, waiting for us to drop. As long as we could hike as fast as they did, as long as we laughed at their raunchy stories and did not take offense, we were safe.

We would wait out the old-timers, we told each other, and then maybe things would change. And there were the good guys, the ones like Dave, who took the time to explain the tactics. He showed me a map of all the divisions and where our line was located. He stabbed a nicotine-stained finger at a cross-hatched line. "That's what is left before we tie it in," he said. He showed me the smooth black line that meant that part of the fire was contained. "Here's our spike," he said, pointing to a number on the map.

"You're doing it all wrong," Dave said another day as we climbed the mountain. "Try this." We trudged up an endless angled slope, en route to a helispot to pick up supplies. In a break with protocol, the crew boss had turned us loose from our usual single-file line. Dave and I rushed to be the first to arrive, Dave because he always wanted to be first and me because I always had something to prove.

Dave called it the mountain walk. Instead of hiking as most people did, ball of foot touching the slope, heel in the air, you pressed the entire foot down into the earth. This gave your leg a chance to rest. "You can walk all day and not get tired," he said. "Steady goin', all day long." I tried it and he was right.

Going downhill in steep country, there was even a song: "Feet apart, legs spread wide, pulaski on down side," Dave bellowed. "My knees are talking!" he exclaimed. We slouched down toward the fire, taking long, sliding steps in the talus.

Sidehilling, hiking parallel to the slope, feet gripping a shifting sea of rock, was another lesson Dave taught me. "Lean away from the mountain," Dave instructed, even though every instinct told me to do the opposite. He pointed at my pulaski. "Hold it on the downward side, so you can throw it away from you if you fall."

We sang the song again: "Feet apart, legs spread wide, pulaski on down side." Though this walk made us look like cavemen lurching along, it kept our center of gravity close to the ground and made us more stable.

It was a strange world I entered, a glimpse into the way men behaved when reduced to the common denominator. I was alternately repulsed and fascinated. Their conversations ranged from the profane to the sublime, one of the trail crew Dans holding forth on Nietzsche while the others discussed the consistency of their morning dumps. They were constantly in motion, using rocks for softballs, lobbing their hard hats down a hill to see whose ended up at the bottom the fastest. They seemed to fit seamlessly into this life. There was no second-guessing, not even a thought of failure. Or so it seemed.

The men grew scraggly beards and chewed tobacco. They carried clear plastic Coke bottles, the contents a vile murky brown. The smell permeated everything, a foul blanket in the vehicles, the work centers, the bunkhouses. Roger stuffed a lump of Red Man into his jaw, working it. His bottom teeth were coated with a gritty brown substance. Steve spit into a trashcan, the floor of his pickup. On the fireline, the ovals of the cans created permanent white rings in their back pockets. Packs were stuffed with big rolls of tobacco to ward off withdrawal at spike camps.

Sometimes after a shift, our feet bare and dusted with talcum powder, the women, if there were more than just me on the crew, let loose our hair and our guard. We talked about female things, things that the guys wouldn't understand. Some of the other women said that they wanted kids, but they didn't know how to be a mother and a firefighter both. Most of the overhead positions were taken by men, who faced no such wrenching choice. It seemed that if we wanted to continue, we had to swing a pulaski forever and be alone in the world. Braiding their hair again, the women on my crew talked about the men at home who didn't want them to fight fire. Their men were jealous of them living in such close proximity to other guys. They wanted their women home where they belonged, in normal jobs where they were home every night. The women painted it as a choice. To keep the men, they had to stop. "Maybe after this fire season," they said, sighing.

I thought that these women loved fire for the same reasons I did. There were times on the line, carrying a tool, when I felt the kind of power that I thought women could never attain in real life. Back there, I felt judged on how much space I took up in the world and how well I fit into conventional boundaries. How many kids did I have, was I married, wait, you don't want kids? What's wrong with you? Meet the right man, you'll change your mind, heh heh heh. On the line life was pared down to the essential bones, the clean slice of a pulaski into dirt, my biceps new and unfamiliar, the nod the men gave me when I held the line instead of dashing out into clearer air, smoke stinging its way deep into my lungs.

I noticed that our small handful of women on this Wyoming fire were all the same age. We had the same stories. Though we were the second generation of female firefighters, none of the older ones were still out here, at least not on our fires. We didn't know where they had gone. They seemed to have vanished somewhere into the smoke. Though we saw plenty of gray-haired men in the upper echelons, there were no older women. We wished we had someone to emulate, but there seemed to be nobody out here but us.

Nobody really knew when the first woman stepped on the fireline. I had heard it was 1970, only nineteen years before this Wyoming fire. It might have been Alaska, maybe in California, but the mists of history obscured her face. There might have been other women before her, but nobody had recorded that story. Eight years before this fire, in 1981, a woman had first stepped out of a plane with a parachute, the first female smokejumper. We whispered about her in the camps, wondering what it had been like for her to be the first. I had caught glimpses of other women on hotshot crews, so I knew that slowly, women were trickling in like water, changing firefighting for good. But there were still places like this, vastly outnumbered by men, watching our backs.

From what I had gathered, the first women were test cases, gaining a trial on a crew because they wanted it badly enough. "If I worked out," one woman told me much later, "they said they would hire more women. If not, that was it. No more." I had heard stories of women being told the only way to get a shower was to go in with the men. Things were better now.

Still, we all felt as if we were embarking on some kind of a test, without knowing how to pass. Above all, we didn't want to be aligned with any criers. When a stranger showed up on our crews, we all eyed her with suspicion. We looked for telltale cracks like ineffectual tool swinging, gear left behind for others to carry, places where she would be weak. I realized that our expectations were higher for our fellow women than they ever would be for a man.

Badger, a wild-haired guy with big stories of fires he had conquered, refused to grid, to stay parallel with the others, but got away with it; other men slouched at the end of the line, doing nothing. They were called out for it, but not in the same way a woman would be. A woman carried the weight of her whole gender. It wasn't fair, but it was just the way it was.

One November in Georgia a woman died, running from a fire. She died wrapped in her fire shelter on a dirt road, the first female wildland firefighter to die in recent history. There was a hushed gasp at the news, then a scramble for justification. Out of shape, no red card, the wrong boots, the wrong place to deploy a shelter. Much was made of the fact that she was female, as if she died because of that.

Our crew was sent to that fire, because fatality fires always triggered additional orders, whether more crews were needed or not. I hiked self-consciously down the line, feeling eyes on me as I passed. It seemed as though all the men watched with narrowed eyes. Would I be the next to break?

In a typical year, about twenty firefighters died. Trees fell without warning, crushing people underneath. Burning limbs flew like darts toward the sawyers standing underneath. Heat crept through veins and smothered hearts. Firefighters dozed off at the wheel. Their vehicles tumbled into space off steep logging roads. Or, less often, the fire got them.

Usually it didn't. We told ourselves that we were smarter, luckier than those who fell. *We would never do that. That would never happen to us.* The reality that we refused to acknowledge was that you could do everything right and still die. Or you could do everything wrong and live. Over the years the fire researchers came up with computer programs that analyzed wind speed, temperature, fuel moisture, and other variables. The programs spit out predictions of where the fire would run and where it would falter. Sometimes they were right.

We took fire classes, hours of staring at charts and puzzling over scenarios. We memorized the fire orders, a set of ten commandments developed to keep us safe. We practiced deploying our fire shelters in less than twenty seconds, giggling slightly as we lay in them, waiting for the crew boss to mimic a wind and try to shake us out.

None of this mattered to the fire. It didn't matter whether we were women or men. It didn't matter whether someone waited at home for us, or whether we were too young to die. We were only fuel, our clothes and bodies and hair no different than leaves, grass, or a fallen log.

It made no difference. Fire was an equalizer. If we didn't learn it on this fire, we might on the next.

We all burned the same.

Size-up

Whoa, there they are," Doug said. "Stump humpers. I've only seen these suckers on fires. And usually near the end of the fire, when things get cool. Check this shit out." We leaned on our tools and stared at the needle-nosed flying insects hovering in a black cloud near ground level. Buzzing like helicopters, they probed burnt stumps with their slender snouts for some reason they only knew. It was just another mystery, something that happened only on the fireline.

That morning I had been walking through grass painted white with early August frost, a skim of snow dusting the high peaks. This always happened so early in central Idaho, where the air was thin and summer as short as a breath. Fires had not really crossed my mind as I reached the trailhead, finishing up a wilderness ranger patrol. The thunderstorm of a couple days before had come with a dense curtain of heavy rain, enough that I thought it would be what we called a season-ending event, no more fire starts. But I was wrong. I had only been back at the bunkhouse a few hours before being called out on this small fire.

When you fought fire, this was what happened. You could wake up in your bunkhouse bed and fall asleep in a different state entirely, with a stranger in the sleeping bag next to you. This would be the person you might have to trust with your life if things went wrong. With each fire, I sized up the new faces: *How much do you know? Will you watch my back?* I was sure they sized me up too.

This time I was with Doug, and we were bound together by five years of smoke and flame. I could relax a little, let down my guard. I had nothing to prove, unlike most of the time, when I was sent out with strangers.

With strangers, the first days were tentative, taking everyone's measure. The crew looked to see what kind of boots I wore: the cheaper Redwings showed less commitment than the handcrafted White's. Even the type of green Nomex pants and yellow shirts we wore told our status. The people who wore the softer, "old-style" shirts where buttons were hidden under the flap and the pants that had deeper front pockets were judged to be more serious than the rookies who showed up in the newer styles. But it was in the work itself that we became a tribe, tied together with an invisible thread. We began to speak the same language, of fires that had been and those that would come, people we had known and had yet to know. As former strangers slung their tools, we carried on an endless conversation, dropped during the times we had to bust butt and picked up again when things got slow. Our conversations were a kind of shorthand that only we could decipher.

"You worked at Big Cypress, down by the Glades? Did you know Mike Patten?"

"The General! Sure did. He's still jumping. Just stopped by to see him in Missoula."

"Weren't you on that helispot by the Lochsa River?"

"Did you fly in on the 212? Yeah, that was a sweet location, wasn't it?"

"Hey, were you on the Yellowstone fires?"

"Wasn't everyone there?"

"Doug, remember that fire in Nevada where we had to carry our red bags five miles because they wouldn't waste a helicopter on two rookies?"

"No kidding, that was epic!"

This time with just the two of us, there were no new stories to tell. As Doug ripped open an MRE and puzzled over a flat package reading CAKE, POUND, LEMON FLAVOR, I thought that all fires were different, yet they were all the same. We had hiked up to this one, compasses in hand, following our maps and our instinct. The chase was always the best part, stopping to smell the mountain air for traces of fresh smoke, the fire invisible under the tree canopy. Before we got there, the fire could be anything—one dead snag, smoking harmlessly on a cliff; an underground labyrinth of roots slowly smoldering; or something bigger and exciting, a running fire that would climb its way up into the treetops and crown them with bright flame. You just never knew until you got there.

After that the real work began.

The first thing we'd done, like on every fire, was to size it up. Rushing in without a plan could kill someone. Instead I hung back, making the fire's acquaintance. Since my first fire assignment, I had shadowed a dozen crew bosses: in the alpine highlands far above the desert floor, in the drought-stressed lodgepole pines of Wyoming and the sequoias of California, learning how to watch the way a fire moved through the trees for clues to how it would burn. I remembered all this as Doug and I hiked the perimeter, taking mental notes on places where the fire would need to borrow strength from the wind to cross: rock slides, clumps of greener grasses. We planned our attack, staring up at the clouds to figure out the weather ahead.

We looked for an anchor point, something we could tie our line into. A rock cliff or a road was bombproof; from there crews could split and work the flanks until they reached the head, the place that was the most dangerous. Without a good anchor point, the fire could creep behind us, trapping us behind two walls of flame. Here, we didn't have a lot to work with, but our fire was so small we could circle it, the two of us, without a lot of effort. We would just keep eyes on it, we told each other, make sure we scratched in a line that we could improve later.

There was a lot to think about during size-up. Could engines reach the fire and drag hose to it? Were there houses nearby? Half-drunk locals in tank tops trying to help? Was there a road where we could make a stand? We had none of that here, only a drought-stressed forest, our feet sunk deep in a carpet of crunchy pine needles, a half-day hike to reach this place.

"Where's your safety zone?" Doug asked, testing me. We went to find one, a place where we could sit it out if things went wrong. A dozer track was good, or a succulent meadow of short, damp grasses, where a fire could sputter and die out. There was nothing like that high on this mountain. Instead, I found a boulder field below the fire that would work. It was unlikely that we would ever need it, but it was the rule.

A safety zone was no good if there wasn't an escape route leading to it. An uphill escape route didn't work if it was too steep because the fire could catch us before we got there. One that was too far away wouldn't help at all. There shouldn't be thick understory to bash through or logs to climb over. That would slow a crew down, and often minutes were all that we had. Doug and I took colorful plastic streamers and flagged our escape route. The woods took on the appearance of Christmas trees, with different colored flagging hanging at odd intervals.

Part of sizing up a fire was looking for barriers like cliffs that could stop its advance. We took advantage of these, herding the fire there with our lines. Sometimes we created our own barriers by burning out fuel between us and the fire, though we wouldn't need to with this small fire. Sometimes we had to make concessions, falling back to places we could defend and letting the fire have the rest. "Bonus acres," Doug said, winking.

All fires had their own shapes and this one was no exception. Ours was a ragged circle, feathering out in uneven points like fingers. Nestled under the tree canopy, it had failed to grow, chewing its way slowly through the forest floor instead. A fire pushed hard by the wind was long and skinny, while one in less of a breeze sprawled out, taking its time. Fires in the steeps fanned out through canyons and narrowed up when they reached constricted spots. Some sent out firebrands a mile in advance of the front, and others spotted across highways and rivers like chickenpox. Others poked around in brush, not moving much.

Sizing up each fire was like learning about a new person, each one with its own characteristics and quirks. Like strangers, the fire could sometimes fool us. Sometimes we thought that we had one all figured out and it snuck past us and was gone. Sometimes we called them dead out and a single spark lingered, reigniting a week later. Being a firefighter taught me to look deeper, at what might be under the surface.

Learning to size up fires meant thinking in a different way. It meant looking ahead into an uncertain future. You gathered what clues you had and made a decision. I often wavered, unable to choose. Each choice could mean that somebody got trapped. The fire could break out of its lines and burn down a town. The consequences seemed too severe to decide.

I knew that sooner or later you had to make a choice. Sitting in warm ash waiting for the fire to decide for you was not an option. Sometimes the choice was wrong. Once Doug and I had left a hard-won fire in the hands of a couple of people, both of us sure that it dozed safely in the fallen timber. Doug was the Incident Commander and I was shadowing him. In this role, I was supposed to make decisions, Doug only stepping in if I was about to screw up. I stared out over the blackened expanse, looking for something to prove me wrong, the telltale signs I had been taught to see. Tendrils of flame licking the base of unburnt trees, the red-hot heart of a smoldering log too close to the line. I saw nothing except lazy smoke rising from a hundred hot spots well within the interior, the separate strands combining into one heavy gray blanket, blotting out the sky.

"A simple mop-up job," Doug told the upturned rookie faces as we prepared to head away to size up another small fire. A cold front roared through that night, sparking it back to life, sending it miles uphill and starting the whole process over, scattering rookies all over the mountain. From our own tiny fire we heard the radio traffic, voices with a hint of panic as the fire sprawled over the mountain. "We're picking you up and bringing you back," Dispatch told us, and we returned to a fire that had gulped up half a mountain in our absence. Mistakes like that happened, Doug said, because fire was an irreverent thing, not bound by our predictions. It could still surprise you, even when you thought you had it all figured out. I filed that away for future reference. This time, all that had happened was the loss of a few timbered acres and the strain on an already tired crew. We had been lucky.

Other times it was more straightforward. If we were fortunate, a helicopter was available and we hiked uphill as it hovered over a thin line of smoke to find a patchwork fire creeping through a stand of pines. The wind was calm, the sky clear. Those fires were easy. We sized them up in minutes and circled them without problems. It was hard to know what you were going to get. You had to go in with all you had and adapt to the situation. You had to give up some control.

Lightning fires burrowed in deep, stray embers we called "duffers" smoldering slowly under the soil. Most often these fires blinked out without anyone knowing; it took a wide expanse of wind and heat to send smoke spiraling into the air. We were supposed to stay on these for twenty-four hours after the last smoke was spotted and extinguished, so we brought cards and books and dozed through the long afternoons. Little brown birds swooped in to pick over the seeds we had churned up and butterflies roved the ashes for trace elements.

The fires I loved best were the ones when there were just two of us, working a small blaze on a sleepy Idaho afternoon. I loved the first sight of the fire after working my way up to it for hours, the uneven beat of my heart as I sized up what was to come. I loved the two of us working in tandem, dirt flying in a fine spray through thick summer air, our three-foot line carving through the layers of soil and needles.

This was one of those, the fires that made me think I could do this forever. "You can take it," Doug had told me back at the ranger station, meaning that I could be the Incident Commander.

The IC! I had dreamed of this for years, working toward it through hours of shadowing crew bosses on larger fires, squatting on my heels watching them draw maps in the dirt with a stick. Even though there would only be two of us, no crew, even though the fire was reported to be less than an acre, I wanted this. I saw from Doug's nod that he thought I was ready.

Doug and I had come up with a plan together. "What about a cup trench over here?" he suggested, and I nodded, knowing that we needed to dig deeper into the slope to catch burning material that could roll past an ordinary line and start fire below us. Some ICs went it alone, their way or the highway, but it was safer to use more than one set of eyes. Sometimes even a rookie could see something I had missed.

Earlier we had called in a helicopter to make water drops, backing far enough away from our fireline so we wouldn't be hit as the pilot hovered low, rotors like a heartbeat in the otherwise still air. I signaled with a mirror so he could find us in the unbroken line of beetle-killed trees, and gave him directions: "Drop it on the north end of the fire, see the big pile of trees that's smoking there?" A shower of water cascaded down. After the helicopter went back for a refill, we sprang up to mix the evaporating water with the smokes, stirring a muddy stew of hot dirt, pine cones, and duff.

Once we had no more work for the helicopter, we started cold-trailing, tracing the fingers of the fire with our bare hands to see if any sparks remained. We slid down slopes on our shovels, returning with piss pumps full of water from faraway lakes. We scooped up shovelfuls of burning logs and piled them up so they would burn out completely. When finally there was nothing left to do but wait it out, we sat on our packs and drank in the silence.

"How many people do you think ever have come up here?" I asked.

Doug shrugged. He was not given to introspection. We were on some nameless mountain, far from any trail. There was nothing exceptional about this particular spot. It was just another lodgepole pine forest halfway up a slope. Through the trees, we could see far below us to the sage-brush-choked valley floor. A river glinted in the sunshine, curving lazily. A cloak of drift smoke from fires bigger than ours draped over the Sawtooth Mountains to our west but I could see their outlines, ten thousand feet and higher. Now that the helicopter was long gone there was only the faint buzz of insects as they hovered over the cooling ashes.

"I bet we're the first people to ever see this," I said.

"No doubt."

"Cool."

"Living the dream."

We watched as the sun winked below the smudged peaks and shadows filled the valley. As the light faded from the sky, we snuggled deep in our bags close to a small campfire we had built inside the fire perimeter. A few stray embers we had missed in our mop-up leisurely blinked out as the cold settled in. The smoke cleared out and overhead more stars than I had ever imagined crowded the sky as if they were fighting for space.

After what could have been an hour or five, we roused ourselves for night shift. We had not been instructed to stay up all night, but we felt the weight of responsibility on our shoulders. If we lost this fire, we lost our credibility. If other crews had to come in after us, mopping up our mistake, we would never live it down. People talked; they did not forget.

Fire at night was different than during the day. It could be two things: slow-moving, embers glowing in the night, or it could be terrifying, an animal crashing through the forest. We passed each other on the line like ghosts in the night. Our bulky shadows were nameless. Sunk in a tiredness that went all the way to the bone, we forced ourselves to keep moving.

Time slowed down on night shift, way down. The hours were sluggish, like pushing through mud. I turned on my headlamp and looked at my watch again. Only five minutes had ticked past. It was always that way.

Finally, impossibly, the distant peaks started to turn the color of chalk. Their jagged outlines pierced a slate sky. Ragged scraps of cloud scooted by overhead, pushed by an unfelt wind. There would be lightning, maybe, before afternoon.

I leaned on my shovel. It was a kind of tired that I had never known before I started fighting fire, even back in my obsessive running days. It was a tired that clogged up my entire body, a tired that coated my muscles, making each step slow motion. Seeing Doug stumble down from the south end of the fire, I pushed through it and met him halfway. We looked over our personal piece of real estate, so far back in the mountains, so far off trail that nobody would ever come here. In the grayness of first light, the fire looked quiet, smoke holding below the soil because of the coolness and humidity rise. To the uninitiated, the fire would appear to be out. Both of us knew better.

As the heat came on, the fire, sleeping underground, unseen, awakened. It was a waiting game, us and the fire. Just when we thought it must be out, that it was an impossibility for any heat to be holding, we found it curled in the arc of a stump face or burrowed deep in a blackened log. We worked one-gloved, our cold-trailing hands grimy beyond recognition, old people's hands, each line marked in charcoal.

Distant fires still smudged the horizon, all bigger than ours. We lay against our packs and speculated where we would be sent next. Perhaps it would be nowhere; rains could come, lightning could strike only the bare peaks. Our bosses could lay down the law and make us stay home.

Besides our morning check-ins, nobody called us on our hand-held radios. Our fire was a pin in a map at the dispatch office, not large enough to rate urgency. The helicopters had been diverted to more stubborn blazes; we were on our own. For a moment I thought that we would be stranded up here for weeks, the fire refusing to die, curls of light gray smoke puffing from the endless heat buried deep underground.

But all fires go out eventually. We reluctantly shouldered our packs and hiked downhill toward our pickup. It was slow going, crawling over stacks of dead logs and traversing around cliffs. I pointed toward Banner Summit, a faint orange glow outlining its location. We had heard snatches of conversation over the radio from that repeater and it looked like a big fire was burning over there in the bug-killed timber, acres of dead trees casualties of the relentless march of the mountain pine beetle.

"Maybe we'll get to go there," I said.

"Maybe."

We were filthy, our clothes and bodies coated in ash that was so stubborn and slimy that it would take multiple showers to erase. We had sucked in enough smoke that we coughed like old men as we walked. These were full-body coughs, raking up and down our spines, doubling us over. They were coughs that seemed endless but produced nothing.

The parking lot was a small oasis of gravel in the forest. Trails spiraled out from here, but went nowhere near our fire. The Pavement Puppy, our low-rider white Ford pickup, squatted alone. It was nearly dark, the hikers long gone to their campsites and motels. Before climbing in, I paused to stare back at where we had been. I imagined that

I could pick out the patch of red-needled trees that signaled it, but I was not sure. Like Banner, bark beetle was marching through this landscape and what I thought was our fire could just be a casualty of the ongoing war. A mere acre, our fire had been swallowed up by the rest of the country.

Back at the ranger station, Doug and I slid into our own pickups.

"See ya on the big one," he hollered and peeled off in a different direction. For three days, our lives had been braided together. We leaned on each other like crutches, unraveling the puzzle that was a fire. We watched each other's backs. Now, just like that, it was over. We might never end up on another fire together again. We might never even see each other again except in passing.

While I sized up fires, I also sized up my life.

This was harder to do. There was no map that I could draw that made sense. Though I worked other seasonal jobs, fire not always one of them, fire was what I kept coming back to. I could barely remember my life before fire. That slender girl, that obsession, the darkness hovering around the edges of my life—I could not go back there.

I sized up who I was.

My body was becoming a map of where I had been, a story etched into my skin. At night in the sleeping bag I charted its geography. It was similar to the topographic maps I relied on to find my way: a puzzle, a series of lines that added up to one real thing.

The mountain valleys where our fires burned were born from volcanoes and ground down by ice and water. My body was the same, a woman who had spent most of her life outdoors and on the move, carved away by sun, wind, and fire. Every scar had a tale, every rainy-day ache a memory.

I did not look like the models I saw in magazines, their skin blank white canvas. My face had been scribbled on by sun. My knees were unstable. In wet weather they sometimes gave a little. They creaked and groaned like an old house. My feet would never win any contests either. Calloused rock-hard, missing toenails and all, they were the feet of a woman well-traveled.

I had learned to read maps. I could tell you where the country steepens up. I could point out where the rivers widened out, sprawled in lazy circles, and became easy to cross. I could show you the grasslands, the deserts, the mountains by reading the contour lines. I had left the girl I was years ago, but there were times when I thought she would follow me forever. I could not drop her on the steep mountain hikes or lose her in the smoke. She dogged my footsteps like a shadow. The only thing to do was keep moving.

It took several fires before I started to absorb what fire had to teach me into my skin. At first it was just a whirlwind of smoke and flame. I followed blindly along, a tool grasped firmly in my hand, and dug where someone told me to. Lined up to eat when someone else said so. Slept in the dirt when I was told. I gave up my freedom in one big gulp. Some of us liked it, others chafed under it. Those were the ones who disappeared at breaks, forcing us to chase them down when the bus came to pick us up. They freelanced, strolling ahead of the crew, picking their own way down the mountains. They were loose cannons and were reprimanded by the crew boss. Sometimes they were even sent home.

On fires, I learned how to chop slices out of a standing burning tree, sleek and slender as potato chips. On my first try I flailed ineffectually and the crew boss laughed and told a guy to show me how it's done. "This isn't your day job, is it?" he cackled, putting on a show for the other men watching. I slunk away, fuming, but sooner or later I got it. Just like that, my hand and arm and tool aligned to strike deep into a tree's core. There was nobody around to celebrate this moment, but I smiled to myself and kept going.

I learned to butterfly stinking wet hose, my arms so short that I was wrapped up in one-inch, my chin resting on the top of the pile. I figured out the intricacies of how to set out a hose lay, the combination of pumps and different-sized hoses that were needed to plumb a fire. I learned which end was the female and what an adapter was and which way the handles lay that meant it was open and which way meant closed. I learned to bleed the lines and determine pressure. I did what a rookie always did, turned on the Forrester nozzle full blast and staggered from the weight of the water, nearly falling. I learned to mist, fine droplets of water, and to straight stream full-force and when each was effective.

I ventured close to a helicopter and crawled under its belly to attach the cables that ran out to a water bucket. I loaded monosyllabic hotshot crews into the seats, buckling their belts for them, adding up their weight and subtracting that from the helicopter's allowable payload. I sat on farmer's fields the country over, watching helicopters of all types rise and land on missions across the fire. I built cargo nets, weighing and settling the needed items—boxes of tools, cubies, saw oil—into the bottom of a wide-mouthed net and pulling it like a purse string to attach it to a swivel. I imagined firefighters like me waiting to receive these things, on a fireline in some forest miles away.

Now, a handful of years in, it was hard to talk about the pieces that made up what seemed simple: the act of stopping a fire. It wasn't simple at all. On rare visits to the Midwest where I had grown up, my friends hollered, "Here comes the smokejumper!"—falling from the sky being the only kind of wildland firefighting they knew. It was pointless to explain the difference.

I knew what they wanted to hear and I told them about digging hotline, building line right next to the flames in a frenzied and direct battle, the back of my shirt as hot as it would ever be without bursting in fire. This was a quick two-step, shuffling bent-kneed as fast as I could go. The fate of the fire, it seemed, swung in the balance of how much I could do with a tool. I told them that the makeup of your crew mattered. There were always slackers, the ones who hung back with the shovels, pretending to need more water than they could possibly hold, faking a blister or two. These were the ones who were in it for the money alone, not the desire. There were the ones who worked themselves into exhaustion, crossing that invisible line. Then there were the ones I clicked with instantly, a glance or a wink all we needed to communicate. If I was lucky enough, my motley crew would coalesce into a long line of something beautiful, everyone moving in unison, the sawyers and the swampers, the lead pulaskis and the shovel-bearers.

I could talk about mopping up fires, listening to the dull clink of my tool against rock, the hiss of steam as I pushed the nozzle of my piss pump forward. How I made a slurry of wet dirt to drown out pockets of small fire and chinked away into the raw flesh of a downed log. Mop-up was the scrape of my pulaski against blackened trees and the hauling of wet, dirty fire hose. My legs turned black too, and my lungs and my face. It was possible to mop up the entire twenty-one days of a fire assignment and sometimes I did. There was gridding too, lining up parallel in a long row, staring at my feet, searching for smoke.

Sometimes, if the night was long and the friend wanted to listen, I could mention the times when I was stretched to the limit, climbing the endless yellow hills near the Mexican border in the waterless desert of Southern California, my squad a sullen bunch of Florida boys, out of shape and out of sorts with having a woman for a boss. When the other squad boss, a man, spouted the same orders I had just given, they hopped to obey. When I had said the same thing, they had pretended not to hear, or smirked outright. On that same fire, we had bedded down in a city park in Cabazon, circled all night by locals driving low-riders. A convict crew was assigned to that fire, and even though they kept to themselves, the rules dictated that women in fire camp could go nowhere alone, even to the porta-potties. A male crewmember was forced to accompany me everywhere, standing whistling tunelessly outside of the johns, waiting for me to finish up.

But most of my friends didn't want to hear that kind of thing. So instead I talked about the times when the delicate balance of wind and flame and air aligned to create a living being, the dance of fire through the tops of trees entrancing and terrifying all at once. The way the breath left my lungs when I saw an entire mountain turn orange. The pulsing red glow of a sun touched by air filled with smoke.

It was not rocket science, firefighting. It was not an art, although it was a creative act to know how weather and topography and fuel came together to create a firestorm. It was an art to know how the fire would move through flashy grass or dense brush or open tree stands. It was an art to know when to make a stand or when to back down. Those were all more feeling than science, although science tried to make it accurate. But the actual act, the getting in there with the saw and the pulaski and the shovel, that was just work, your body in a motion that could be sustained for hours, days.

It was not even glorious, although the rookies always thought so, and it sometimes seemed like it when we drove down from the mountains and the town had put up signs saying THANK YOU FIREFIGHTERS, the same kind of signs I had seen on my first fire. It sometimes seemed like it in the airports when people walked up to me, people I didn't know, and told me how brave I was. If you were a rookie, this made you smile and think differently of yourself. If this wasn't your first rodeo, if you had been around for a while, you would know that it wasn't really about bravery. You might wonder just what it was that kept you doing this job, because if you said it was about the money you might as well quit. No, it was something else, something that was hard to put into words. When everything went right, it was almost like poetry, a rhythmic line of yellow shirts under hard hats, one fluid movement of dancers, and fire paralleling them through the smoky trees.

It was years before I started to think about whether all this firefighting was really good for the forest, if it might not be better just to let it burn. The first time I thought this, I watched the California fire service bulldozers cut a wide path through thick chaparral. Moments before a convict crew had barely escaped a burnover, rushing to the road and taking shelter under an engine. A helicopter had flown too low in the smoke and become entangled in power lines, landing in desperation at a staging area, crews scrambling for cover. Looking up the hillside, I saw faint white lines where other bulldozers, other years, had bladed the same firelines. It all suddenly seemed pointless, a fight against nature that we could not win. I thought this, but I sure as hell didn't say it. It seemed like a betrayal to even think it. To believe that what we were doing was pointless was not something I wanted to think about.

I saw how we left scars wherever we went, overturned earth and raw stumps and sometimes entire firelines that the fire never reached. The tankers sprayed slurry indiscriminately, sometimes getting it into forbidden streams. I saw firefighters throw spent batteries into the flames and leave toilet paper blossoms just off the line. We definitely left a trace.

There was nobody I could talk to about this. Everyone else seemed so united in the pursuit of fire. Get it, catch it, stop it. The idea that we could let some of these fires burn was hardly ever mentioned. The idea that fire was a natural force and could be good for the forest was never mentioned either.

As I began to learn fire, I also learned the subtle differences in the trees. I had never thought so much about forests before. I began to see them as fuel. At first it seemed to be all the same, flames and smoke and dust, but as I traveled around the country I saw that fire was shaped by both terrain and fuel.

Most of our fires were in lodgepole forest, because that was what blanketed a large portion of the West. Lodgepole was made to burn if only we would let it. Its cones were serotinous, the resin gluing them shut, and the heat of fire could cause this bond to break. Lodgepole fires could be stand-replacing, an entire forest gone in hours. Nowhere was that more apparent than the granddaddy of them all, Yellowstone in 1988. This was the fire that became legendary, the result of what happened when nearly a century of fire exclusion and a diminished snowpack collided. Almost 800,000 acres burned as several small fires were pushed by wind to combine into monsters.

I had gone to Yellowstone that year not as a firefighter but as part of a crew conducting research. We walked through forests still hot from the burn, laying out transects, long strips of measuring tape at randomly determined locations, and collecting soil along those lines to determine how much of a seed bank was left. We wore hard hats and Nomex and carried pulaskis, but instead of fighting fire we were studying what it left behind.

Other types of trees burned that year in Yellowstone, but it was the lodgepole that I noticed. In the unburnt sections of forest, places the fire had inexplicably skipped over, our four-person research crew scuffed through calf-deep pine needles, an acidic barrier to any other vegetation. These forests were dark and the canopy formed a wall through which the sun could not shine. I saw that these forests needed to burn, but most of the time we would not let them. I saw the consequences of this as I marched through Yellowstone, tool in hand: huge jackpots of downed trees, brushy trees with limbs we called ladder fuels. They were a time bomb waiting to go off.

Anything would burn if it got hot and dry enough, and I went on after Yellowstone to fight fires in sub-alpine fir, high on the mountainsides. These were places recently free of snow and I felt the bite of winter as we patrolled the line. Sub-alpine fir was often killed by fire; it lacked natural defenses such as thick bark and roots buried deep in the soil. Its neighbor the whitebark pine had the same problems, except that the whitebark had an alliance with the Clark's nutcracker, which would sometimes bury seeds in open, burnt areas and forget to come back for them. As I walked the fireline I heard the rasping call of these birds punctuating the minutes.

I learned to fear fire in the ponderosas. Even though their thick, wrinkled bark made them more fire-resistant than most trees, slow-moving fire insinuated itself in their bases, making scars we called cat-faces. Without a warning, the tree could topple over, instant death for whoever stood below.

Chaparral in Southern California was a killer too, head-high and flammable as spilt gasoline. We couldn't go in there if it was burning hot. Manzanita crackled and spit and carried fire if it was dry enough, but regerminated from seeds dormant in the soil after a burn. Even the desert sagebrush had its own intensity, firewhirls like miniature tornadoes spinning across the flats, unstoppable and treacherous.

I learned that if we didn't mess with the cycle, fire and the forest had a long relationship. For some fuels, like the Florida swamps, fires visited the same areas every three to five years. In other forests, it could be hundreds of years before the right combination of aging trees, disease, and lightning aligned. Seeds waited patiently, deep in the soil, for generations. Fires ripped spaces in the closed canopy, sunshine streaming in for the first time. Everything in the forest was poised for fire if we would only let the fires happen.

Instead, we went after them.

There were things we used to stack our odds against a fire. I learned all of them, the language of water and air. With a big fire, it was generally too late for retardant. Instead the overhead team pulled out the big guns, the Vertols and the Sikorskies, the heavy helicopters that could carry thousands of gallons of water at a time. They fought fire with fire, arming a crew with drip torches and letting a backfire rip across hundreds of acres, counting on the delicate balance of wind and heat to cancel out both fires. I shot small flare guns loaded with flammable shells deep into the woods to generate spot fires. I asked for and received gifts from the sky: pumpkin-shaped water holders called blivets that I filled using portable pumps, nets full of bladder bags and hoses and hose fittings. I chose my weapons: pulaskis for harsh, rocky soil; rake-like scrapers called McLeods for fine fuels like pine needles; flappers and wet burlap sacks for grass, just like I remembered the veterans telling me long ago. I even used a leaf blower in Tennessee.

And then sometimes we waited. There were fires so volatile that all we could do was wait them out until they calmed down. We sat on the school bus, bored to the point where people started to catch and eat bugs for money, while the crew boss paced and listened to the radio. We went through all of our jokes and our card games and our lies. Someone convinced someone else to shave his head into a Mohawk. Some read books. Others worried the sharp point of their pulaskis with flat files until the shavings coated their pant legs silver. This was the time that separated those who were good at this from those who were not. The ones who had no patience were the ones who would not make it in this game.

"Equal parts boredom and terror," it was said, and it was learning the boredom part that was the most challenging. It was letting go the thought of yourself as a hero, armed only with a shovel. It was knowing that in the end, weather and topography had more to do with putting a fire out than you did.

Some never got it. They strutted, believing the hype. It was true that we helped, especially in the case of small fires. There it was get in, line it, get out. But the big fires like Yellowstone in 1988? Snow put that one out as winter came to the high country. We didn't make much of a dent in it, except to herd it to natural barriers and keep it out of Old Faithful. Back then, in 1988, we all believed that Yellowstone was an anomaly, a two-hundred-year event. Surely there would be nothing like it in our lifetimes ever again.

We soon would know that we were wrong. We would see it over and over again as the West warmed. Fire seasons were pushed back weeks at a time. On a tiny blaze high in Nevada, the crew boss shook his head. "Fire in the alpine fir, this early," he said. "Never seen it before, not like this."

I learned that we were slaves to the weather. It was a capricious bitch, sometimes good to us, sometimes not. Some years you could wring out the air like a mop, too much rain and too little heat to get any fires going. Others boiled and fried but were so hot and dry that thunderclouds could not form. It was impossible to know what awaited us, though we tried to predict it, staring up at the snowpack each winter.

Lightning was what we all wanted, the god we worshipped. People started fires too, but we had a bias against that method of firestarting. It seemed like cheating. Instead, we wanted fire by natural cause. We wanted fires that started as one flash of lightning, forking out of a slate sky in some inaccessible, lonely place. We wanted the spark to burrow deep into the duff, slowly channeling its way through the layers of needles dry as bone. We wanted it to travel silently on the parched ground, reaching a crispy-leaved bush or two, to pad soft-footed, skunking around in the understory for a few days, gathering its strength. After that a solitary wind could come, blowing up canyon to the peaks and ridges, enough of a breeze to get some momentum behind the fire, enough to push it into the trees.

In our minds we built the perfect fire, just like I dreamed of the perfect man, the one that real ones never quite measured up to, the one who let me go but took me back when I came home. The perfect fire would be easy digging but plenty of overtime, a crew that gelled with no slackers, huge flames but no life-threatening danger. The real fires came close but never quite got there. As good as they sometimes were, I thought they could be better. They were just good enough to keep me hoping for more.

We all had our lightning stories. Doug and I were nearly hit on a ridge, bracketed by fiery strands that touched down on all sides as we cowered in a forest of long-dead trees. One night at fire camp Little Mike and I watched under a yellow tarp as a fire blossomed to life on a sagebrush slope above us. Others told of their hair rising skyward and their pulaskis buzzing with light as a storm passed by.

If we desired lightning, we cursed the rain. A brief and intense downpour touched us less than an all-day soaker, the rain needling its way deep into the hundred-hour fuels, those logs that lost and gained moisture in that length of time. We were doomed in the rain. We sloshed around the fireline, our pants plastered to our ash-covered legs, our boots unredeemable, fat drops bouncing off the rims of our hard hats. Black plastic sprung up around the inadequate tents at camp, and grim rumors of demobilization—demob, or being sent home—flew around the ranks. We were all on the lookout for the dreaded season-ending event, a rain of long duration or an early snowfall. Either could put us out of business and back on the street.

With the others on my crew, I chewed over the forecasts like grizzled bones. La Nina or El Nino? Drought or spring rain? We jealously watched the quadrants of the country. Damn, we cursed, it was torching in the Southwest, while here we sat in Idaho, snow spiraling lazily down in May. Or: This might be the year for Alaska, guys!

After several years, I had carved out a place for myself. I could show up on a fire with my pack and my pulaski and figure out what needed to be done. People listened as I spread out a map on the hood of a pickup and described where we would make a stand. Most often I was one of three squad bosses, the only woman with that responsibility, each of us leading a group of five crewmembers, subordinate only to the crew boss. My squad followed me in single file as I led the way through the forest to the line. "Where's your safety zone?" I asked, testing them. "Where would you run if this all goes to hell?"

When I dragged my Nomex fire-resistant pants up over my hips and buckled my belt, when I buttoned my soft, faded yellow shirt, when I slid my feet into fire boots as soft as butter, I was putting on my own type of armor. I stood taller. I took up space. It was the first time I had ever believed in myself. Even though I never spent a summer on a hotshot crew, never jumped from planes—preferring the freedom of a small do-everything fire crew in Florida and the best of both worlds, wilderness ranger and firefighter, the rest of the year—I thought that the long-ago hotshot woman might have seen me as a sister on the line now. I had earned it.

I knew one thing: I would do anything to stay in the game.

How to Get Over a Smokejumper

Like all firefighter girls, I knew that love, desire, and the fireline were inextricably entwined, a complicated knot impossible to untie. The surge in my belly at the sight of a fat column of smoke eating through tinderbox forest was the same ache as wanting somebody you weren't sure was good for you.

Maybe real love wasn't like that, but I didn't know the difference. I had never been schooled in it. Because I had chosen fire and a life on the road, I ran out of time before I could really know the locals, the men who stayed, the ones with roots down deep. In a season, what could you know of someone except the way they breathed deep in sleep while you lay awake and thought about where you would travel next? Time became a series of fleeting moments, impossible to gather up into something substantial and lasting. I was always out the door to somewhere else and I never knew when I would be back.

Smokejumpers were my weakness. There was something about them that was irresistible. There they were, striding through fire camp in their White's boots, their

Nomex shirts grimy with spilled bar oil and sweat. There they stood in the chow line, trash talking, faces still black from ash. I heard them on the radio, from the roughest places, the most dangerous sections of the fire, while I was stuck cold-trailing on a distant ridge. The best thing to do, I knew, was to stay far away. After all, I had heard all the stories and the jokes.

What's a smokejumper's birth control? His personality.

All smokejumpers get divorced.

Remember the five-hundred-mile rule: when a smokejumper is five hundred miles from home, anything goes.

I heard all this and knew there had to be some truth in it, but I fell anyway. I sat in a tiny bar in the Sierra Nevada mountains, watching a line of sweat-shirted firefighters sprawled on stools, still smelling of smoke, their eyes little fires in their angular faces. These were the most alive men I had ever seen.

It was last call and then time to go. The whole group stood outside for a minute, the early spring chill burrowing into our bones. Wasn't it natural to want some warmth after a long winter, the radiant heat of someone who spoke the same language of fire as I did? When one of the men, the quiet, blue-eyed one with close-cropped hair, looked right at me, I felt known. Here was someone who got it, someone who understood how fire pulled at my heart.

"Can I come over?" he asked. "I've got booze." It was late, well past midnight. I didn't drink much, not like these guys, who seemed fueled by a steady diet of alcohol and flame. I didn't need a man; though I was not immune to the touch of someone's hand on my summer-dry skin, I was more in love with mountains than with men.

"I guess so," I said, sensing danger but unable to resist. As we sat at my kitchen table, drinking liquor straight up, I felt lost already, a crack in my armor widening until I could no longer patch it back up. I sat back, arms folded, taking him in.

There was something about him that made me want to let down my guard. I was known by then for loving men and letting them go, fish back to the sea. I wasn't sure I wanted to do that with him.

He was lean and spare, extra meat and fat trimmed down to the bone. Most of the time, he seemed poised to strike, like a snake, holding back raw energy. Part of a hard-partying, slim-hipped bunch, he was different than the slightly effeminate, tie-clad national park rangers in their shiny shoes. Those men were books I had been through; I knew the ending already. There was no surprise with them. They wandered potlucks snapping up the leftovers, braying nervously at their own jokes. They parted their hair in the middle and showed up for work in ironed pants. There was no surprise with them. The man I chose, on the other hand, was rough around the edges. He wasn't one who would sit around and talk about feelings. Instead he was on the move, darting from pull-up bar to shower, bar stool to pickup truck. He was seldom seen, a challenge, like the mountains I climbed on my days off, like the fires I scouted. When we finally kissed, I could taste the cloying sweetness of Kahlua on his lips and something else I had never felt before. Maybe everyone else had it right and I had been wrong. Maybe love was like fire, a lightning bolt warming the air hotter than a living person could stand, frightening but exhilarating at once. Maybe it could last, for once, past a season.

The West was thirsty that summer. Cumulus clouds billowed thousands of feet overhead, fires going nuclear. Dry lightning every afternoon brought small wisps of flame slowly working their way to the tops of trees before blooming into life. Crews from other parks were bused in, a yellow-shirted army loitering around the parking lots waiting to be sent out. The radio crackled with orders for hose and mop-up kits. Our sleepy enclave was transformed to a war zone, helicopters darting overhead, lead planes flying low and slow, tankers in their wake.

Sometimes they called for him when it was barely light, wanting an early start. Often I thought that I had dreamed it, sitting wrapped in blankets, watching him lace up his boots. I tried to think of something to say that would make a difference, which would stick with him while he was gone. *Keep a foot in the black? Watch your back?* Even something I could not take back, something that opened me up in a way I never had let happen before: *nobody will ever love you the way I do, come back safe and alive to me?* But I didn't say it.

We revolved around each other in a sort of limbo, kept apart by fire. Weeks later, he slipped back through my cabin door, bringing the smell of wood smoke into my bed. He told me places he had been: other states, other forests, night shift, it all blurred together. Even while he lay beside me, I knew he would rather still be out there.

There were only four hundred smokejumpers in the United States. In order to become one, you had to pass grueling tests of your mental and physical toughness. In other words, you had to have the right stuff—a quality that was instantly recognizable but not definable. Either you had it or you didn't. Dozens washed out every year, sent packing because they didn't measure up.

On the crews I went out with, there were always one or two who wanted it badly enough. Sitting on their packs, waiting for the bus, they mapped out their future. This handcrew stuff for a couple of seasons, enough to work their way up to lead pulaski or crew boss. Then a few seasons on a hotshot crew. After that, jumping. Their eyes turned hazy. This was the ultimate.

"You should come with me to Florida. There are fire jobs down there," the smokejumper told me as we prepared to leave our summer jobs in the national park. The people who were really serious about fire went there, and he told me that you could learn more on a crew there than you could in twenty fire seasons out west. Down in the Glades you got to do everything—pilot swamp buggies with engines mounted on the back, direct helicopters over the wet cypress and dry prairies, your feet sinking into a deep stew of mud, boss around bulldozers and crews on fires all your own. Everything burned down there, he said, everything. It burned hot, crazy hot, unlike anything I had ever seen. His eyes radiated the kind of excitement he reserved for fire only.

The tourists disappeared from the park like turning off a switch. The snow line traveled farther down each day. Waterlines were being drained, other seasonal workers heading out for good. I knew I couldn't stay there. I couldn't let him go, but I hadn't managed to secure a winter job yet, and he would be lost in my hometown with its manicured lawns and high expectations. My parents would be horrified by his Copenhagen habit.

I had heard the Florida fire stories and he told me more: fires that made their own weather, fires that skunked around in the swamp for days, forcing him to chase them through poison ivy and water that sloshed over his boots. There was more fire there than I would see in a lifetime. He waited for my answer.

I thought about it for a few more minutes. Winter had the West in a chokehold, our summer park jobs days from termination. The brave few stayed on to eke out an existence volunteering or squatting in remote, waterless cabins, but most of us fled east and south for winter seasonal jobs. Real firefighters, the people who couldn't let go of their addiction, headed east to the Everglades or Big Cypress, parks sunk deep in the Florida swamps, where you could have all the fire you wanted. You could have fire started by stealthy turkey hunters, wanting to produce a carpet of juicy green grass shoots to seduce their prey. There was fire started by red-neck arsonists who just wanted to see the woods burn in a firestorm of volatile oils from the flammable vegetation. Then there was fire we started ourselves to reduce wildfire risk or to clear out the brush for the elusive, endangered Florida panthers that were making their last doomed stand there.

I had tried giving up the seasonal life a year earlier, heading back to Michigan for the winter. Low-slung clouds formed a sullen barrier against the sun. I had forgotten how flat the Midwest was, the endless horizon unbroken by mountains. I worked at an airport, a job meant to be permanent, collecting weather data on the night shift and advising pilots of the weather on the ground. I was the only person in the building on those long nights, the runway lights burning holes in the darkness. No planes landed and I fought against sleep, wondering if this was what life was supposed to be all about: clocking in, putting in eight hours, heading home for a restless day sleeping. I recorded the many types of snow on a government form: graupel, sleet, pellet, snow showers, mixed rain and snow. I squinted up at the cloud cover, which could be obscured, overcast, broken. There were so many words for absence of sun.

The day workers drifted in as dawn broke, their faces as gray as the sky. They spoke in terms of decades. Forty years and out. Like the flat country, it seemed as if they could see all the way to the end of their lives. Winter was a constant; it would never be spring. I put in my two weeks' notice after a few months.

I said yes to Florida, following a man across the country, something I had thought I would never do. We took our separate lives and separate pickup trucks through the endless expanse of Texas, the lush low country of Louisiana, and finally weaving our way through the interstates and forever drained swamps of the Sunshine State.

The man I followed, I learned too late, was angelic when he was happy, his hair a pale halo, his eyes kind, but anger could boil out of him with one wrong thing said, and I said the wrong thing all the time. He fought not with fists but with words. I had never fought with a lover before, but his words were needle-sharp, mostly in response to the way I clung too tightly, trying to pry me loose. They pierced deep into places that were still soft and I lashed out to keep from crying. Our fights were cataclysmic, punctuated by the calmer days when we took our drip torches and our flare guns and our fusees, all tools to make fire, and set a prairie alight. We were always at our best when the day included fire because it was something we could agree on. The first puff of smoke, the twin lines of a flanking fire, the unrestrained joy of a wind-tossed conflagration: this was a language we both wanted to speak. His face under his hard hat, lit up by a backfire we had started, made me burn with a love that was often unreturned. Most weeks he strolled in and out of my life, claiming confusion. The reality was that he loved fire more than he loved me, and would leave me for it, though I did not know it then.

We burned hundreds of acres at a time, dancing on the balance of wind and humidity. The smoke column rose higher in the sky than a plane, creating its own thundercloud. We saved ramshackle hunting camps by burning the fringes around them, carefully creating a blackened edge where the main fire would not cross. The saws whined as the chainsaw blades bit into the grassy hearts of the cabbage palm. I threw out rebar sticks, measuring the time it took for flames to move between them. I took the weather, spinning a damp cotton wick attached to a sling psychrometer, making sure it was still safe for us to be out there. I recorded all this information in a little notebook and turned it in to the crew boss, who added it up for an official report demanded by the overhead.

Above us the helicopter was a fly in the sky, buzzing inconsequentially. Where we were was the real show. We labored down the muddy trail in the swamp buggy, an oversize vehicle that looked like a cross between a tank and a jeep. We took our long-handled flapper tools and smothered errant flames that threatened to creep across our line. We worked late into the night, sometimes all night.

I lived in a bunkhouse with three other women and at night cockroaches ran across our faces. The smokejumper had dangled the promise of living together in his single room, but when I arrived, carrying boxes to his door, he snatched the promise back. He met me at the doorway. "I'm not ready to live with someone," he said, finality in his tone.

Curious occupants of the building peeked out of their own doors. What had he told them about me? I felt foolish, standing out there with all of my possessions. "I wouldn't have moved here alone!" I hissed. "You said . . ." I began. His brows drew together. Anger like storm clouds began to gather in his voice.

"They can hear us fighting!" I whispered, but he shrugged, preparing to close the door. "Lighten up!" he said loudly, and I heard a few snickers from behind closing doors. I knew what he meant was that he wanted a party girl, someone tough enough to fight fire but who didn't demand anything other than an unending stream of nights without promises. I had never been able to be that girl and I couldn't do it now.

He slammed the door.

Saying he wasn't ready meant he might be someday, I rationalized, loading the boxes back into my car. He had left his heart open just a sliver and I thought I could worm my way in, over time, if I were good enough, perfect enough. The women in the bunkhouse made room for me. They didn't say that they had told me so, although I saw it in their faces.

"Maybe we should just be friends," he said often, his face furrowed with doubt. "Maybe," I said. After he had left for the night, I curled into a ball on my grimy kitchen floor, my body pulsing with pain. The other women made themselves scarce; this was something they both understood and wanted to distance themselves from, their own memories too painful.

Always he came back, squatting on his heels, asking for another chance. As soon as I said yes, he changed his mind. "I just don't know what I want," he said, slamming a fist into his palm. He stormed out the door. I never knew when he would come back, but I knew he would come back. I knew I wasn't strong enough to let go.

We boomeranged between love and hate, unable to choose. The days got hotter, the rains came, and people started talking about summer. The smokejumper was noncommittal. None of the possible futures he mentioned included me. I had left men before, but I wanted to be done with leaving. I wanted someone who could make me stay. I thought this man might be the one.

In my bunkhouse, there were maggots in our rice and bright green geckos on the walls. We were constantly out of money, unsure of our futures, and just plain afraid, but fire made up for it. When we were working on a fire, the adrenaline running through our veins, we were completely alive, glowing from within.

"Why are you with him?" one of the other women asked as we lifted weights in the seasonal workout room. She shook her head in exasperation. She, like the others on the crew, was collateral damage to our boomerang love. They were uneasy witnesses to our back-and-forth wrangling, our raised voices. Deep down I knew that we were a liability to the crew, our heightened emotions affecting our actions. Her boyfriend, a firefighter also, was calm and steady. They planned where they would work in the summer together. They never fought, at least not the way we did.

I had just watched a slender blonde bob in the smokejumper's wake, following him like so many other women did, and I banged the weights with more force than necessary. Despite the big weight stack I pulled down to my chest, I felt fragile. I stayed in the weight room for hours at a time, wrenching heavy dumbbells over my head, willing myself stronger. "You're almost lifting heavier than I am," he said, pausing at the gym door. A smile briefly flickered across his face. "Got to go," he said, pushing himself away from the door. I could count the times he said he loved me on one hand.

Our fights were always along the same fault lines. The smokejumper wanted the same freedom he wanted in the air, the freedom to steer away from anything that would entangle him. Trees. Power lines. A boulder field that would break his bones. Me.

There were two types of parachutes that jumpers used: square and round. The round ones had a static line that pulled the ripcord automatically. Square ones had to be piloted from the start, the jumper pulling his own cord at a specified altitude. Either way, jumpers experienced a few precious moments of freefall each time they stepped out of a plane. These few seconds, while they were falling unhindered, transformed them. When they came back to earth, they were forever changed. My smokejumper's eyes were far away, full of clouds. I reached out to try to stop him with protestations of love, but he slipped through my fingers every time. Nothing I did could compare to stepping out of a DC-3, feet first into fire.

At the end of the season, he was as ephemeral as smoke. I had imagined that we would live the seasonal life together, a seamless flow of winter and summer without end. We would never have to grow up, get real jobs, and become too old to fight fire. "I don't really know what I want," he said again. He didn't ask me where I wanted to go. He meant to continue on the road without any ties. The more I tugged, the more he pulled away.

I didn't recognize myself in this woman, burning with a kind of fever I couldn't name. I blamed it on the rootless years that had come before, the desire to have one person who waited for me to come home, the one phone call from fire camp that meant something. He didn't get it, but the men never did. They thought that when they wanted something different, it would be easy to find.

In the end he went to Idaho to jump out of planes, and I grabbed a job as a seasonal wilderness ranger in the same state. I could still work fires, and I would get to build trails and hike solo across the backcountry. The fact that he would live as far across the state as he could was tempered by the fact that I had always wanted to work as a ranger, a plum job as far as seasonal positions went. "Do you know how to use a pulaski?" my future boss asked me. "You need to know how to use a pulaski." Yes, I told him, I did know.

It was going to be a good season, the smokejumper said. The snowpack was down and there were already lightning fires, even this early. "Lots of oats," he said, meaning overtime. Overtime meant that he could work twenty-one-day hitches with only a couple of days off in between, not long enough, I knew, to drive across the state. I wouldn't see him for months.

I knew the seductive force that was fire. It stood between us like a third person, a love triangle I forever would battle. When a fire broke out, his eyes lit up in the way I wished they would for me. I wanted him to love me as much as he did fire, but fire always won.

I stood by his pickup waiting for him to leave. It felt like goodbye. The air was heavy with unshed moisture, warm air punching through with no hope for a cool breeze. The wet season was starting. Soon the rains would cover all the buggy trails with four feet of water. Lightning would dart the slash pines but any fire that was left would go out in a hurry. We both knew it was time to go.

A few days ago Andy, one of our crewmates, had thrown his fire boots over the power line, declaring himself finished with fire forever. They dangled from a knotted lace, swaying slightly in the wind. Others had gone before him, saying they were going back to school or getting real jobs. How could they walk away so easily? *That won't be us. It will never be us.*

"I'll call you," the smokejumper said. He swung into his truck and drove away. I watched the brake lights tap once, as if he were changing his mind, but then he passed out of sight. We were headed to the same state, but it felt like it might as well be across an ocean.

We both ended up in Idaho for the summer, but we rarely spoke. Limited by pay phones, neither of us were easy to reach. I had been on a few fires that summer, but it was a slow season after all, the snowpack lingering into July. Now it seemed as though the light was being drawn down, a tide moving out. The sun hung lower in the sky. Waterdogs, leftover fog from a brief rain shower, snuck through the trees. The forest breathed. If I didn't know better, I would think it was smoke.

We broke up over the phone. From the wilderness work center barn, where the phone hung, I stared out over the Sawtooth mountain range to the west, absently picking out the peaks as I listened. Thompson, where a woman had cartwheeled from the summit that summer. Mount Cramer, with a lake at its base where a mountain lion had stalked my campsite one July night. Many more unnamed summits known only by their elevation. In the growing dusk a faint pink blush spread across the range, the last of the sun we would see for that day. Alpenglow was always the saddest time, when the bright possibility of the day gave way to the night.

Across the yard, the trail crew Dans traded a bottle of Jägermeister back and forth. One light-haired, the other dark, they sat like bookends on the bunkhouse steps. The taller Dan waved the bottle at me, an invitation to come drink with them. "That's not your personal phone!" he reminded me. There was a three-minute rule for us and I was well over. I turned my back. The conversation was strained, unfamiliar. "Why didn't you tell me you met someone else?" I asked, an echo in my head reminding me that others had asked me a version of the same question: *why do you have to keep moving on? Why can't you stay here?*

On the steps, my roommate Deb had joined the two Dans. She must have said something that they found especially silly, because the taller Dan snorted and said what he always did.

"*Jesus,* Debra."

Normally this made me laugh, but I was waiting for an answer.

"I didn't have the guts," the smokejumper finally admitted, though I thought that jumping out of a plane into smoke took more courage than to say goodbye. The old words, the ones I had kept at bay for so long, came flooding back. *Not good enough. Not perfect enough.* I had thought myself free of that whisper inside my head, but here it was, back again.

Later, I heard through the fire grapevine that the smokejumper ended up marrying the woman he had chosen over me. "Yep, we met her that summer. We went to dinner with her," a firefighter friend told me a year later, averting his eyes from my stony glare. His seesaw between the two of us had been well documented, yet I was the last to know.

Deb and the Dans watched curiously as I slammed the phone back in the cradle and brushed past them. They didn't follow. We all had our casualties of the seasonal life. They had been there too.

The smokejumper sent me a letter after that, one sentence written in pencil on lined notebook paper: *I'm sorry, but this is the best for both of us.* On the envelope, he had misspelled my last name.

Nearly a decade later I was on a small lightning fire in the mountains. Jumpers had gone in the night before to scratch a line around it, and as usual, our crew was sent in the next morning as the mop-up squad, the jumpers off to better things. Shouldering a saw, bladder bags, and pulaskis, we threaded our way uphill through lodgepole pines. The lazy smell of smoke drifted down the mountain and we could see slow puffs of dirty gray, a sign that not much was burning anymore, perhaps just some downed logs.

This would be one of the good ones, I thought, a small fire far away from any overhead, just dig and shovel and put it out and go home, earning good overtime and hazard pay in the meantime. On these fires, we were cut loose from the burden and tyranny of fire camp. These were the fires that really worked the way they were supposed to.

"Jumpers," someone whispered. They came walking down the hill, a line of five in faded yellow shirts. It was early enough that the sun caught the pollen in the air, turning it into gold. My smokejumper was with them, second from the front. Our lines passed close enough to touch. We looked at each other under our hard hats, smiled a little. I felt the old tug between us. It was still there even after many years apart. There was so much I could have said. Then again, there was nothing to say. Maybe if we had led a different life, not as firefighters, the ending would have been the same. It was impossible to know.

On a fire I had to focus on nothing but the fire itself. Get distracted, that was when things went wrong. I saw this all the time in men going through divorces or mired in sticky custody battles. The rest of us had to watch them closely because they weren't squared away. Their minds drifted away from safety zones and escape routes. They had one foot out of the black, always. It was a dangerous state to be in. Already the rest of the crew was far ahead; I had hesitated as the jumpers passed. I had to close the gap. I hurried to catch up, the moment lost.

Since there were no creeks near the fire, we dry-mopped it, turning over the soil again and again, revealing the pine cones and leaves that smoldered deep in the soil. We spread hot rocks out to cool and crack in the mountain air. The boys peed on the hottest spots, giggling at the steam.

The fire was out before dark, a random black patch in an unremarkable forest. We made our way down the steep, dry slope, coming out at the guard station and our pickups. The next day a skeleton crew would hike up to check it, running ash-covered fingers through each inch of dirt to make sure it was truly cool.

My crew chattered as they stowed the tools in the bed of our six-pack. "Let's go into town," Justin said. "We can get pizza and beer at Bob Dog's."

"Or go to a movie," someone else chimed in.

I barely listened. The sky was empty; the jumpers had folded up their chutes and left a long time ago. I felt sort of lonely, standing there in my thin shirt. I knew that when I got back to the bunkhouse, my roommate Deb would be gone out with the trail crew for ten days. Everyone else was scattered through the wilderness, on fires or on projects. It was only August, weeks to get through until I packed up the truck and left again. What I had not bargained for was the emptiness I felt after a fire was over, after we had rolled all the hoses and sharpened all the tools and gone through our packs to throw out the remains of old lunches. I lived from fire to fire, mountain to mountain. For the first time I had the treacherous thought that this was not enough.

Justin leaned far out of the pickup like a dog, his shaggy hair blowing in the wind. Tina palmed the wheel, our truck tires rattling over the washboards, a rooster tail of dust showing where we had been. Their faces showed nothing but weariness and contentment. Was I the only one who wrestled with my heart, stay or go? Was I the only one who wrestled with my conscience, walking through Yellowstone in 1988, the unburnt forest a biological desert, the burnt parts with a vibrant seed bank waiting to spring free? *I don't know what I want,* the smokejumper had said so many times. I realized that I finally understood what he was trying to say. There was so much to know; it seemed like it would take a thousand lifetimes to learn it all. There were so many paths to take, and no way to know if you were on the right one. You just rushed headlong on the one you had picked, trusting that everything would work out.

Justin chattered about the next fire. There had to be one, silently creeping through the drought-stricken pines, a backcountry smoker, just waiting for discovery. This was the year, he proclaimed, meager snowpack, pine-beetle kill. The residents of the valley had protested loudly against any kind of prescribed burning. They didn't want blackened trees in their view. Outgunned, the Forest Service gave up, instead blanketing each fire with a rapid response, even the one-tree blazes sequestered high on rocky slopes. I knew that this was making it worse, that someday whole hillsides would go up in a firestorm.

We fell silent as we drove down valley, the red corpses of dead trees on the moraines slowly advancing more and more each year. Big fires were coming. While I wanted to be part of the excitement of the inevitable siege, I knew that the fire we had just suppressed would have harmlessly cleared out unhealthy brush and puttered around in the rocks before slowly winking out. We had not been necessary at all.

Walking Through

Roger and I pushed through the Florida swamp, trying to stay on our separate compass bearings. I could tell where he was by the crashing sounds through the knee-high palmetto. On the other side, Steve was only a suggestion, faint noises of his passage the only sign as he was forced farther away by accident or terrain. Jen and Mike paralleled us, lost somewhere in the thick growth of Unit 5. We were walking through to the other side.

We were deep in the heart of the state, flanked by the glitter of Miami some sixty miles to the east, the booming town of Naples to our west. The wildlife refuge we worked in lay just off the interstate that belted the bottom half of Florida, separated from speeding cars by a chain link fence. This was a little pocket of land that had somehow been saved as a wildlife refuge, a distant memory of what the southern half of the state had once been. It was a last gasp of wilderness in a place that had long forgotten what it was to be wild.

The property of a former hunting club, it had somehow escaped the steady march of pavement and golf courses on either side. Barely thirty thousand acres, the wildlife refuge burst with life. Everywhere I looked, there were dozens of shades of green. Muhlenbergia grass grew exuberantly in huge tussocks like teased hair. Bushy cabbage palm crowded in, shouldering through the skinny trunks of slash pine. Saw-leafed palmetto burst from the ground, forming dense thickets. I thought sometimes I could almost hear the plants growing, a constant background hum. The forest's breath was palpable, all the competing plants drawing great gulps of air as they struggled to find room for themselves.

In winter, the Florida fire season, water drained from the land, exposing sun-baked mud flats. In the summer, the refuge was awash in fresh water, tinged brown from tannin and sloshing against the silver skin of the cypress trees.

We used the rock-studded jeep trails as boundaries for fire units. Jed was itching to burn the unit Roger and I walked through, but endangered birds were thought to lurk within, and we were dispatched in this unscientific manner to find them. It was the only way to conduct a search in such an overgrown place. Flying over wouldn't do it. Even used to the place as we were, we sometimes got lost, staring at aerial photos and trying to pick out the clearing we were in. Without landmarks, we circled until we found something we recognized, or Mike climbed a tree, looking to spot familiar ground. Sometimes I thought that I could wander off into the swamp and never be found, even though the refuge was held fast by the roads that bordered it.

We were supposed to stay fifty feet apart from each other so that the entire unit was surveyed for red-cockaded woodpeckers, but ponds and thick vines kept pushing us on the same line. I glanced over to see Roger in his bright yellow shirt, straying into my quadrant.

"You're supposed to be over there!" I yelled.

"Did you see the wild pigs?" he screamed back. Out here on my own, cutting through an unfamiliar world of musky-smelling water and sharp-sided grass, it was comforting having him so close and I found myself listening for the sound of his muffled curses as he forced his way through the thick stew of a living swamp. He was keeping track of me too. We tied pink flagging onto our baseball caps and sometimes all I could see was the bright ribbon, bobbing along through palmetto.

I was single during this dry season, my heart one deep bruise from where the smokejumper had left it. Like the older brother I had never had, Roger was slowly showing me how to heal. We never talked about the smokejumper, even though Roger had been the one to tell me that the jumper's heart had belonged to someone else all along. Bound by the brother code, nobody had told me until it was too late. I didn't blame Roger. All of the guys protected each other.

We talked about what we saw instead. Candy-striped snakes that could be either deadly or harmless depending on where the stripes lay on their thick bodies. "Red on black, friend of Jack," we chanted. Scarlet king snake. Then, "red on yellow will hurt a fellow," the poisonous coral snake, a pit viper that would latch on and not let go. We talked about the alligator nests we found and the whimpering calls of the tiny babies as we waded through waist-deep water with cans of boundary marking paint. The sweet smoky taste of heart of cabbage palm, impossible to resist even though we knew that harvesting it meant killing the plant. The mystery of why one orange tree, growing feral out on the refuge, could bear such sweet fruit when its neighbor's was mouth-puckering sour.

The first time I met Roger, someone was dragging him out of his room clothed only in underwear for the temerity of going to bed instead of joining a raucous party. A smokejumper in the summer, he came down to extend his fire season in winter here. Still in his underwear, he stood up and accepted a beer. The sound of his laughter filled the room, so infectious that it spread through the room like a virus.

Though I had grown up only a hundred miles from Roger, we had never crossed paths until now. We shared a common north-country heritage, making Cornish pasties that our grandmothers had once baked, flattening out the buttery dough with a wine bottle in lieu of a rolling pin, chopping up potatoes and rutabagas to stuff inside. We snuck through the work center intent on practical jokes on the rest of the crew: mousetraps in slippers, Vaseline on locker door handles. When I couldn't finish a pull-up or back up the trailer, Roger never cackled loudly like the rest of the guys and showed off his own prowess. Instead he gave me pointers. *Let me hold your ankles while you try again. Turn the wheels this way. You've got this.*

"Riiiiiicolaaaaa," Roger yelled, imitating the throat lozenge commercial. "Sorry!" I called. I had drifted too close to him and I corrected, cutting back onto my bearing.

Roger and Jen and I had been sharing a spare FEMA trailer since the beginning of the season. These trailers were meant for hurricane survivors and Jed had obtained them through his combination of southern charm and insistence. The trailer was parked a short stroll from the main work center building, and it became the lunch spot for everyone. Bill even clomped up the steps every noon, to watch *Jeopardy!* on our small TV.

A single-wide, the trailer possessed little charm. Our rooms were three cubbyholes, carpeted in nubby green. We fumbled around each other in the tiny kitchen. The building swayed with every footfall and every voice echoed. There were no secrets in trailer living.

By now I had lived with closeted drunks, a kid who claimed he could walk on water if only he believed enough, and giggly interns who pounded on African drums and smoked weed. Roger and Jen leaned into each other like plants looking for a common sun. They spoke the secret language of couples. I was fascinated by their certainty in their relationship. How did they know it would last? With a string of failed loves behind me, I was no longer certain of anything.

My bearing took me into a saw palmetto patch that stood taller than I was. This unit had not burned for decades and brush was forcing the deer out. Like us, they struggled through the pricker vines and the dense understory. No deer, no panthers. It was a simple equation that we tried to change with fire.

Saw-toothed leaves nearly blocked out the sky. I clutched my compass tightly in one hand, eyes fixed on the slender red needle. It was easier to get lost in a place this flat, no canyons or rivers to navigate by. Each prairie looked like the one before it, each hardwood hammock similar to the last.

Here in the deep woods it was easy to believe that the state did not have a wound at its heart, that panthers were not inching closer to extinction with every cat hit and killed on the highway. In here I could forget that our refuge was an island in a sea of development, golf courses and high-rises and more and more people, enough people that it seemed the peninsula would sink with their weight.

The only time that the town residents really thought about nature at all was on the days when fire roared into their neighborhoods. They built their houses inside webs of vegetation so thick that it was impossible to save them, but we tried, dragging hoses and torches, fighting fire with fire or water, whichever tool in our arsenal worked. Part of the reason we burned these units was to create fire breaks so those other kinds of fires would not happen, but we could not burn everything; every year more acres torched, taking houses down to the concrete foundations.

I pushed through the palmetto, breaking out into an open pine stand. This was the place where the woodpeckers might be, if indeed they were here. I wanted to find one, for the novelty of it all, so I circled each tree, craning my neck upward for telltale signs of sap and holes. Unlike other woodpeckers, these birds nested in living trees much like the ones I walked through now, trunks gnarled and twisted with age, thick boughs frozen in an eerie dance. It was hard to find these kinds of trees anymore. Many had disappeared, victims of hurricanes and wildfire. Larry the biologist had told us that 97 percent of the bird's habitat was irrevocably gone. This was why we walked through each unit.

"A bird just flew by. I'm going to follow it and see what it is," Roger called on the radio.

"I'm in a big pine stand," I reported. A series of answering radio calls echoed through the unit.

"Fording a big pond," Jen said, out of breath as she waded.

"I'm almost to the buggy, you slowpokes," Mike responded. "How long before you get out?" Steve, waiting for us, wanted to know. Of course, deep in the unit, we had no idea. It could be a hundred feet or a hundred miles for all we could tell.

Poison ivy draped the trees in a solid mat. Strangler figs had them in a chokehold too, slowly advancing until they took over the tree to reach sunlight. Bulbous cigar orchids, now dormant, hung from branches. The whole swamp bristled with independent life. As I walked I summed up what would burn and what would not. This dark pond embracing shaggy cypress trees would not burn; the fire would part ways and move around it. The clumps of wax myrtle would crisp up as the flames crackled through. Our burns always aimed to create a mosaic, a patchwork of lightly scorched areas in some places, heavily burnt in others, imitating what used to happen in real life before humans got involved to mess things up.

There were no black and white birds, no flowing sap, and I conceded defeat, moving on. Another large pond gaped before me and I stepped carefully around it. Those could be bottomless, the soft mud beneath the water sucking off boots for good. An ancient deer stand perched in a tree above, nailed boards for access. Wild pigs had rooted in here, throwing dirt into large piles. Their musky scent hung in the air.

"Steve, give us a hoot," Roger said on the radio. I stopped and heard Steve's howl, muffled but not too far away. We were almost there. I flew through a pocket prairie, my feet squelching in the wet grass, and fought through the last palmetto field, clawing my way into the open.

Roger and I popped out on the fire trail, only a few feet from each other. He snickered as he took me in: shirt ripped partway down my shoulder, hair a snarled mess. The buggy idled, Steve working a lump of chew in his jaw; Mike was sprawled on the back deck, gnawing on his customary unlit cigar. He had gathered some palm fronds and was in the process of weaving a sun hat. Jen sat shotgun, her hair the color of brown sugar in the light, her camo pants soaked above the knees. They lifted their arms in a lazy wave. We had made it through to the other side.

I liked walking parallel to my crew, separate yet together, bound with an invisible thread. It felt a little less like I was alone in the world. We reeled each other back in when we needed it, making it through lonely nights at the trailer, sitting barefooted on the front steps when we wanted to breathe in the warm, scented air. Frogs croaked out love songs in the small pond behind us. Next door, past the deeper pond and through the cabbage palm trees, two old people maintained an animal farm, filled with cages of former movie stars: black panthers, tigers, and lions. Their growls could be heard from our front steps, long lonely cries of animals behind bars, far from any kind of home they could recognize.

Those were the nights when we told every story we could think of, skirting the edges of vulnerability as close as we dared. Walking unseen through the burn unit, or side by side laying down strips of fire, I could feel our bond expanding over the distance between us, thicker than blood ties had ever been.

"Clear to burn," we reported to Larry, who had been waiting by his radio at the work center. The Student Conservation Association volunteers geared up to descend upon the unit with telemetry antennas, looking for the signals of radio collared Florida panthers. If there were any in there, we would have to hold up until they moved. "I'll call for a burn permit tomorrow," Jed told us over the radio.

Steve started up the buggy and we moved in harmony with its uneven progress over the rough limestone. Bill clattered by on the dozer, cigarette dangling from his mouth, disking a safety line, giving us a wave. We surveyed the unit with practiced eyes, noting the danger spots, the overhanging trees that might throw sparks, the snags that should come out.

A tampon rolled out of the pocket of my cargo pants and Roger scooped it up. "Think you dropped something," he drawled with a wink. I snatched it from his palm.

We arrived back at the Florida Panther work center, a squat white building that had once housed a family, now repurposed into a maze of small rooms stuffed full of ancient metal desks, bins of tools, and a bathroom with an odiferous toilet that often refused to flush. Stinking of tobacco and orange hand cleaner, the place had little to recommend it but the noisy swamp cooler that bellowed icy air into the main room, a respite from the steamy outdoors. We fought each year to hack back the jungle of tropical vegetation that threatened to swallow the work center whole, and for the most part the brush was winning.

We tumbled off the buggy, headed in different directions. "Grease the buggy, get the tools, get the aerial ignition machine ready," Steve ordered, settling a battered ball cap on his head. "Burning tomorrow!" But we already knew what to do; it was a routine we had performed many times. I hustled to the pole barn, a concrete-floored behemoth with a tin roof supported by old telephone poles. Grabbing the grease gun, viscous pink slime oozing out of its tip, I slid a mechanic's roller cart called a creeper underneath, searching for each silver nipple I needed to coat with grease to prevent water damage from reaching the engine. Clumps of wet mud and grass dripped onto my head as Mike above rinsed off the top deck of the buggy with the power washer. My hair got caught in the creeper wheels as always and I yanked it free.

Roger crawled under the buggy. "Missed that one," he said, indicating the point with his own grease gun. We had spent hours together under here, fixing one thing or another. The underside of the buggy was a familiar map. So by now was Roger's face, intent with concentration as he aimed the gun at a point. He stared at the underside of the buggy with a practiced eye, noting rust blossoming on the axles from salt water.

We finished up and shot out on our creepers, flying across the cracked concrete floor of the pole barn. I laughed as I got up from a sprawled position, several strands of hair missing, and wiped my filthy hands on my cargo pants. If anyone from my past had seen me now, they would never recognize me. I had sprouted firm biceps and sun-hardened skin. A long braid hung heavy over one shoulder. I thought back to the woman I had seen so long ago on the line. I hoped that I resembled her now, most of the softness replaced with muscle.

We worked late into the night, the big halogen lights on in the pole barn, bats and nighthawks punching the sky. Steve, Jed, and Bill, who all lived in town, took off for the half-hour drive to Naples while the three of us remained.

Roger fired up the oldies rock station. It was a sort of dance as we wove past each other carrying armloads of tools, as we filled up the pump, as we rolled around on creepers under the vehicles. I could feel the approach of summer in the way the humidity rose in a great steaming breath from the jungle surrounding us. Soon it would be time for the three of us to move on.

We neither celebrated this nor mourned it. After all, we would be back here in six months' time. The intervening time stretched like taffy, unformed and sweet, and we knew that this dance would begin again. Bill would retire, but someone else, younger and ready to be beaten up by the caprock trails, would take his place. New interns would appear bristling with telemetry equipment. But the trails would still fill with water as the summer rains came. The fragile ghost orchids would still bloom; the fat, showy cigar orchids too, though less and less as runoff from the tomato fields up north killed off their main pollinator, a bumblebee. The swallow-tailed kites would dip low over the borrow pit, and the first of us who saw one was promised a six-pack from Larry the biologist. Fires would punch through the abandoned subdivision to the west that we called the Blocks; it was guaranteed. It was something to count on.

In our trailer bunkhouse, Roger gulped from a bottle of gin and labored over a tiramisu that he was determined to get right. We waited in a lineup for the one shower, sweat rising from our hair and clothes.

While we waited, we stared at the world map pasted on the brown paneling above the couch, loudly proclaiming our someday travel plans. This was a continuous, circular discussion that resolved nothing but raised endless possibilities. Some firefighters even got sent to Australia during their burning season, Roger said. Australia! We dreamed about that a little.

We each felt a little uneasy about this dreaming, realizing the selfishness of following our own plans when others expected different things from us. Roger wrestled with it more than the rest of us. His dad owned a tire shop and could use the help, but how to give up all of this? He would jump a couple more seasons, he decided, and see what happened.

Roger commandeered Jen's sewing machine. Last summer he was kicked out of the smokejumper sewing loft in Idaho for not sewing well enough, and he was determined to get back in. He threaded a piece of cloth through the needle, trying to construct a sack for his Thermarest pad.

"What are you doing, Roger?" Jen and I hovered over him, trying to disturb his concentration.

"I'm making a stuff sack, I've come up with a plan," he sang. His bony foot worked the pedal. Jen and I eddied around him, brushing oil into the cracked leather of our boots and setting up our Abs of Steel video.

"Come out and do this workout with us," we urged him.

"I have abs of aluminum," he said but good-naturedly joined us, leaping up and down on one of the wooden boxes he had made us for this purpose, woefully out of step with the music.

He and Jen had just gotten engaged after being together for years. "I'm getting married in my fire boots," Roger grinned, and we joked about elaborate scenarios where the wedding would resemble a fire camp. On our days off the three of us traveled into the swamp, searching for mystery. We hunted up ghost orchids, their tendrils locked firmly onto pond apple trees, blooming only in July or August when we would be long gone. Delicate and endangered, they were hanging on here in the refuge, but we weren't sure how long they would last. Air plants, epiphytes that got all of their food and water from the atmosphere, clung to branches and trunks, making trees look like they wore fur and bristles. As I walked, I collected the sightings like souvenirs: small, willow-rimmed ponds, fat banana spiders stretched over webs, impassable vines I somehow had to crawl over, my body held up by their sheer density. The gumbo-limbo trees, nicknamed "tourist tree" because their peeling red bark reminded someone of sunburnt visitors. Secret stands of silver-barked royal palms, eighty feet high. The midnight-colored water that sloshed around the base of cypress trees. Elaborate deer hunting stands high in the trees, left by the hunting club. We flitted back and forth through the trees like shadows, calling to each other to come see our latest discovery. Since the refuge was closed to the public, we were fortunate explorers, seeing what nobody else got to see.

As I sat in the back of the buggy, the two of them in front, I wondered if I would ever find someone of my own. As close as I felt to Jen and Roger, at night they went into their room, shutting the door. I was alone then. I could hear them laughing and talking in low voices, telling all of the secrets that couples kept from the world, and I sat on the front step of the trailer, smelling the rich stew of growth and decay and waiting for the next fire, the next adventure to make me feel whole.

Close Call

"What did you do today?" the cute cop asked. He scanned the parking lot for trouble. The pavement radiated leftover heat under my unfamiliar heels, worn to impress. The cop's short spiky hair and profile were in shadow, lit only by the green glow of his dashboard. He was only half with me, like always.

"Did the mile and a half run," I said. He nodded, but his eyes were on some teenagers loitering outside the liquor store. A row of sunshine-haired girls lurked in the alley—his fan club, I called them. They were waiting for me to leave so they could swoop in like bats and lean over his window like I did. Hoping to be chosen.

"Roger and I crushed it," I said, but he was listening to his scanner. There was trouble somewhere in the eastern section of the city. His face was serene and inscrutable as he thought through a plan.

Despite myself, despite what I knew, I was trying to turn him into my boyfriend, and some nights it almost worked. The nights when he paged me and wanted me to come over to the immaculate house he lived in, a house that looked like no woman had ever shaved her legs in the tub or brushed her hair over the sink. The nights when fire blossomed out of the Blocks and we exchanged a charged banter over our fire tools. Those were the nights I let myself believe I could have what Roger and Jen had, someone to tangle myself into like a pretzel, someone I could let see my real self, the softness underneath. On those nights Jen and Roger watched me slip out of the trailer in a sundress, my hair hanging down my back, and said nothing, even though they must have known how it all would end. It hardly ever ended well for seasonals and locals. Someone had to give up too much. In Nevada, a girl had stayed, slipping from one life to another like a reckless leap across a chasm. The rest of us watched uneasily as she moved boxes out of the bunkhouse. We had never heard from her again or seen her on the road. It seemed too big of a gap to cross, to stay.

I knew that I was not the girl the cute cop would marry. I was the secret nighttime girl, not one he would hold hands with at the seafood festival or kiss in the street. He knew I was leaving soon; why would it be any other way? I always fell for these guys, the ones who kept their hearts locked up. The easy guys fell for me, the ones with hearts wide open. They tagged after me like puppies, eyes big and brown. I didn't want them one bit; I didn't desire their eager plans for kids and homes, lives far from the fireline.

The cute cop clicked on his seatbelt. "Got to go," he said. "Good to see you." He was lost to me already, even though I still had two months left here. He had never really been mine to lose.

"See you later. Page me," he said as his window rolled up. Half the time he did not answer his pages and I suspected it might be because he had a whole stable of women to call upon, a lineup of Florida blondes sweet as sugar. Or instead, it could be because he was wary of girls like me, the ones who always left.

In the end, it was better to count on fire. There would always be another fire. Men came and went, leaving grooves on my heart, but fire was my constant. All it took was a spark on dry wood. The forests were tinderboxes, hushed and dark as night under a choked canopy. They waited to burn, needed to burn. Little light filtered in on the trails. Tall, spindly lodgepole pines fought for sun in a forest starved for fire. Winter storms had toppled some, and they piled up head-high in huge windthrow just beyond the trail. It was only a matter of time.

In the decades before, western forests were under the "ten o'clock rule"—the goal being all wildfires put to bed by ten the next morning after discovery. This well-intentioned policy of keeping fires small had the effect of allowing suffocating fuel loads to build up. Now we were locked in a vicious cycle: we couldn't let the fires burn because they burned too hot, threatening the crop of new homes that had sprung up in the mountain subdivisions. Left alone, the first fires would have crawled through, not touching the canopies, clearing out the undergrowth. Now it was too late. Our cause was not helped by Smokey Bear. For generations, he had convinced the public that all fire was bad. They now believed it with a vengeance.

In the Southeast, where we spent our winters from November to April, fire burned white hot. It crashed through the spiky palmetto with pent-up rage. It flared up in the slash pines and tore through the prairies. It was beautiful and deadly. You never turned your back on it. We heard the stories: the dozer operator whose machine got high centered, stuck fast on a stump, as he was punching in a line right at the head of the fire. Once Bill scorched an ATV when he parked it next to the fire. Another time Roger had to come to his rescue with a flapper, smothering out the flames as they threatened his stalled machine. We talked about them all: the near escapes, lightning slamming into summer trees. We made a joke of it, most times. Fire would never turn on us, we thought.

If we missed one fire, there would be another. We could count on it like we counted on the sun rising every day and Roger stumbling out of the trailer, untied shoelaces trailing from his boots, his whole body erupting in an old-man cough from years of built-up smoke.

Roger and I ran. We ran as if our lives depended on it. If anyone asked us, we would have said that our lives did indeed depend on it. Without this job Roger would be up north, changing tires. I would be somewhere, but not here, maybe at an airport on the night shift, or having babies in a northern town. We wanted to be here, so we sprinted down the highway toward the far-off pickup where Jed sat with a stopwatch. Our fate hinged on seconds.

The rest of the crew had fallen way back. Though they could all run quicker if they needed to, they saw no point in going any faster than the required time to pass the mile and a half run test. They joked and jostled for position on the banked road. Out in front, it was just us.

Besides Steve, our foreman, Roger and I cared about fire the most. The others liked it, would cheerfully hoist hoses and flappers as the occasion demanded, but they did not love it as much as us.

They carried their pagers reluctantly. "Again?" they moaned as we were called out right at quitting time. "Some of us have lives," they said pointedly. They moved reluctantly toward the buggy to hoist the pump on board. They cast longing glances at their watches. They liked fire, but it wasn't a lifelong love for them. They could see themselves doing something else.

But I loved fire, getting itchy under my skin if a few days went by without it. Without fire I thought that I was just another girl, nothing special about her. In Roger I sensed a kindred spirit. We talked at night in the trailer about all the fires we had fought and all the ones we had yet to see. I leafed through my fattening file of fire stories: the little one high on Cony Peak in the Idaho wilderness, a three-person crew in the middle of a vast untamed place with a pair of shovels and a pulaski, unnamed lakes glinting far below. The fire among the shadows of giant sequoias, my crew dwarfed beneath century-old trees. Chasing smokes in Nevada, high above the desert floor.

A steady stream of traffic lurched past on the two-lane, vans in various states of decrepitude belching black smoke and lacking mufflers, headed for the town of Immokalee. It was a normal route for them, migrant workers heading to the fruit, drug traffic moving north. The drivers rubbernecked, puzzled. They gunned their engines and yelled encouragement, though they must have wondered why a group of five was running down the road as fast as they could.

We ran on the shoulder where the state was widening the road. There was a bone-jarring slant to it that I felt with every step. My shoes were old and beaten down with miles, but I was willing them to survive another season.

We ran on churned-up, uneven dirt, skirting piles of cut and stacked cabbage palm, fronds yellowing as they died. This was more acreage of the original subtropical forest hacked and bladed and drained and subdued beyond recognition. Our refuge boundary skimmed the fence that contained it, an island of lushness in contrast to what lay outside.

Though the clock barely topped nine, sweat soaked my cotton T-shirt. My dank ponytail slapped the back of my neck. The humidity made the air thick and sluggish, as if we were swimming instead of running.

Roger's skinny white legs under long baggy shorts churned up the road and I struggled to keep pace. Whatever happened, I would not let him pull ahead. We did not have enough breath to talk but in an unspoken agreement, we were staying together. I spared a glance over my shoulder: the others were at least a minute behind, maybe two. They slowed, realizing that they didn't have far to go. We had halved the distance between Jed and ourselves; in the distance we could see him, his legs swinging off the tailgate as he waited for us, his floppy cowboy hat pulled down against the sun. He shouted something, but over the sound of my breath and the blood pounding through my ears I couldn't hear a thing.

To keep our jobs, we had to complete this mile and a half run in eleven minutes, forty-five seconds, and we could coast in at that exact time like the others. That was a slow time, easy to make. We wouldn't get any prizes for beating it, but we gave it all we had. Roger and I were alike that way.

I had done this run for almost ten years, in many different states, and this was by far the fastest I had ever run it.

Roger took off fast when Steve told us to go, faster than he usually ran. He glanced back at me with an impish smile, knowing that I could not resist the competition. I was off like a shot, right on his heels.

Though most firefighters now did the step test instead of the run, Jed made us do both. The step test involved stepping up and down on a box accompanied by a metronome. Our pulse afterward indicated our fitness, with a target number of heartbeats that we could not exceed or else we would fail. It wasn't a fair test, some of us muttered, because it favored those with a naturally low heartbeat, regardless of the reason. People on heart medication, determined to fight fire despite their condition, passed it easily. Men and women stepped on different boxes, regardless of height, and the men could have a higher heartbeat and still pass. It was rumored among the women that our scale was harder, developed when there was an active movement to keep us out.

Jed figured that we should pass both tests, so we ran too, lacing up our sneakers each year, although there were rumors that this would soon be supplanted with some other test. There was really no good way to see how people would fare on the fireline without just putting them out there. Some dropped dead anyway, even with the run, felled by some secret ailment. Others just folded for various reasons. Sometimes the most unlikely person would stick with it for years and years, like I had.

Running was nothing new to firefighters. Steve had been a hotshot before coming to our crew, and he liked to make us do a hotshot-type run where we lined up in single file, passing a volleyball over our heads to the person behind us. If one of us dropped the ball, we all had to grind out pushups. But more often we just ran as a group, peer pressure spurring us forward.

In the damp Florida mornings we gathered with bleary eyes outside the work center for our run. It was early, but not early enough to escape the humidity that pressed down like an invisible ceiling. Steve called our runs "PT" for physical therapy, but there wasn't much that was therapeutic about it. It was an all-out war, some of us trying to make the others crack. "How much farther do we have to go?" Mike puffed. "Half mile," I said, lying through my teeth. Discouraged, he slowed and I giggled and kicked the pace up a notch, pulling ahead. Realizing my ruse, Mike couldn't let me get ahead. We raced at high speed until we collapsed, panting, near our pickup.

This was miles away from my lonely, obsessive runs before I started fighting fire. Then I raced only the clock and my thoughts. Each footfall was like a slap: *Not fast enough. Not thin enough. Not good enough.* Now I felt essential, braided in with the others in the crew. The faster and longer we all ran, the more we could protect each other on the fireline. We encouraged each other to make it to the next cabbage palm, fence, abandoned house.

Now the running was paying off as Roger and I closed in on the end. I couldn't see Jed's face but I knew he was grinning at us and our contorted expressions, our form going to hell. He had been through this more times than we could count and he knew what it was like.

Suddenly I hit a wall. My breath came in ragged gasps. It was less than a quarter mile, under two minutes, but the distance to go might as well be miles. The weak part of me, the part I tried to punch down every day, thought: I can't do this. Who cared, anyway? Eleven minutes forty-five was fine.

Roger noticed and for a moment I thought he would sprint ahead and claim victory. That was what I would have done. But he didn't. He dropped back with me. "Come on, we got 'er!" he hollered. This encouragement was all I needed. I coasted in on fumes, nearly falling as I pulled up near Jed and the cooler of water he leaned against.

Jed looked at his stopwatch. "Nine and a half minutes," he said. He nodded, tacit approval. It was not all that fast for a firefighter; the smokejumpers could do it much quicker. It was fast for me, though. Roger stretched and slapped a mosquito. He acted like he had just been out for a leisurely stroll. I watched the others jog up to the finish. Not winded, they laughed and drank water. "Why'd you guys run so fast?" Mike asked. We shrugged. "Because we wanted to," Roger said.

"Load up," Jed said. "Tram brushing today." We slid into the pickup for the short ride home, three to the front, three to the back, our sweaty skin sticking to the slick seats. At the work center we pried ourselves out and, lacking time for a shower, pulled on work-stiffened clothes: camouflage pants, T-shirt steeped in transmission fluid, flat-soled Vietnam-era jungle boots, chosen because the swamp water could run in and out of vent holes, preventing foot rot from trapped moisture. We wore these clothes and boots, acquired from the local army surplus store, instead of the more expensive Nomex when we were doing project work.

"Tram brushing," Mike said in a tone of gloom. He chewed on the stub of an unlit cigar as the rest of us rushed around gathering gear. Somehow Mike was always ready before the rest of us and he watched us hurry with a barely disguised smirk. He always stayed cleaner than us too, and we never figured out why or how.

The tram lines were old raised roads established by loggers lusting after the large cypress trees. They laid rail and scooped out the swamp to build several miles of these paths. Later the rails were removed, but the evidence remained. The oaks and palms crept in over the years, so now the tram lines were like long, leafy tunnels riding high above the swamp. Barely wide enough for a buggy, they stretched in straight lines, east and west and north and south. Owls flapped out like ghosts; sometimes the limbs of trees touched above, blocking out the sky.

The tram lines were a battlefield of opportunity, the trees scrambling for light and rain, the palmetto crowding the understory like crooked teeth in a mouth. Leftover wind from hurricanes punched deep into the corridors, toppling the ancient oaks. Jed wanted the paths kept open as natural fire breaks and we jockeyed with chainsaw and axe to clear them. These were days where the whack of our axes and the growl of the saw went on for hours, the sweat collecting in every fold of our skin. We moved only a few feet in an hour at those times, measuring accomplishment by inches. Scarlet and yellow spiders fell down our shirts and our cutting arms grew unstable. Saddleback caterpillars launched from trees, their midsections a pale green resembling a saddle blanket, brushing their venomous spines against our skin, leaving a deep burning sensation where they had touched. Everything on the tram lines seemed both beautiful and dangerous.

We had lucked out, though. It sounded like we would be burning instead. Used to rapid changes of plans, we all perched on wheeled office chairs to wait. Jed was on the phone with the Division of Forestry. He held the receiver so we could hear the question.

"Where's your contingency line?" the dispatcher wanted to know. I could hear a smile in her voice. Everyone loved Jed, with his deep southern accent and the way he poked fun at just about every situation. I knew a joke was coming as Jed pondered his response. Where was our contingency line? This was the place where we would make our final stand if the fire got away from us.

Jed grinned and winked at us. "Well, we got the Gulf of Mexico," he said.

We all knew it for the joke it was. Our fires always stayed within the lines. They never got away from us. We had it all under control. Nothing could surprise us anymore.

The day that fire turned on us was unremarkable. It was another simple winter day in a series of South Florida dry season days, the sky a sweet forgiving blue. Our seasonal friends were gutting it out in the winter snows, fire only a distant dream. We taunted them with postcards and letters and phone calls. We're going to burn today, we wrote. If you were here we might let you drag a torch. Living the dream, we said.

The wet season was coming again. The rain had begun, not the gentle drops of winter but gushers that wet only one section of pavement, leaving the next mile high and dry. The trails snarled with mud, becoming impassable except by the bravest among us. The air was thick with humidity and promise. With wildfires dormant, it was time to burn again.

Elsewhere in the country prescribed fire was a shifty thing, easily sent galloping into a subdivision by a change in the weather. Fires out west could burn a long time, months even. A single mistake and houses went up in flames. In the South, aided by high evening humidity that winked the fire out by midnight, we could let it rip, burning up thousands of acres in an afternoon. Even though our smoke columns could be seen for miles, residents in town went about their business with casual equanimity. They didn't seem to make any connection between the place they lived, which had been hacked down and drained and beaten into submission, and the wild swamp just a few minutes from their doorsteps.

In the pole barn, Mike lay under the buggy, tools scattered in a silver semicircle. "What do you want, Darted Monkey?" he asked, chuckling a little at the nickname he had given me, when I came to tell him we were burning.

The buggies were always breaking down, casualties of the muddy water and limestone caprock. We spent hours under them troubleshooting.

Mike rolled out on his creeper. "I'm just going to lay under this buggy until someone tells me different," he drawled. But before he could roll back under Jed came out with our orders. Head to Unit 54, double-time.

I thought through the map of the wildlife refuge in my head. Unit 54 was the piece of land way to the west, bound on one side by private land. It lay close to the Ford test track, where Mike and I had slunk through the woods to peer at the prototype cars being driven around by the motor company. We had been up there not long ago, prepping this unit for a burn.

We had gotten ready to burn many times before. It was old hat to all of us and we went through the motions: pull frozen water bottles out of the freezer, top off the buggy's water tank, shrug into our Nomex, still mouthing peanut butter sandwiches. Bill swung up into his dozer with a groan, settling into the seat and clattering ahead of us down the limestone fire trail. Jen, Mike, Roger, and I followed in the buggy, Steve at the wheel, carefully skirting the famed mud holes in which we had all mired buggies in the past. Jed stayed behind, waiting for the helicopter to pick him up.

At the unit, I prepared my drip torch. Mixed just right, the liquid within held a purple tinge. Mixed "too hot"—too much gas, not enough diesel—and flames dripped off the wick and roared up the handle. Not hot enough and nothing would burn.

Steve dripped a little fuel down on a patch of saw palm and I touched my lighter to it. I held my torch close to the fire until a flame worked its way up the wick. We surveyed the scene.

Lines of slender-trunked pine mixed with waist-high prairie grass in this unit, making tactics difficult. The helicopter was already laying down fire in the interior, and our job was to ring the unit with black. I saw Steve thinking it through. We didn't want to send a crescendo of wind-driven fire into the pines, killing them. We wanted to keep the pines and clear out the grass and brush.

Steve spit Copenhagen into the grass. When he was devising a burning pattern, he took his time and thought things through, hands on hips, dark eyes scanning the terrain and sizing it up. He said finally, "Only one burner." I knew that he wanted to keep a slow fire backing its way into the unit instead of our usual triple line of parallel fire, which would burn too hot as it combined. "Walk toward the pines and bring fire with you."

Roger, consigned to the buggy, flipped the switches to turn on the pump and climbed on board. He had decided to grow a beard this winter and endured Steve's nickname of Scruffy for it. He idled the buggy slowly behind me as I lay down fire in a long row. Jen followed on foot, snuffing out any embers that snuck across the buggy trail. Behind me, an even row of flame marched toward the horizon. As usual, my whole life seemed to click into place. It didn't matter that it was Saturday, everyone else in this state floating bare-toed down the Peace River or lying under a cloudless sky at Lover's Key. It didn't matter that somewhere in a place shrouded in winter the smokejumper had spoken marriage vows to the woman he chose over me. It didn't matter that the cop was doing a slow fade as summer approached. Nothing mattered but the chew of fire through the clumps of muhlenbergia grass, the satisfying black line, starting out narrow and growing as it burned back into the unit. At times like these the gap between the woman I used to be and the one I was aiming for narrowed and became invisible.

It was like dreaming, walking with sun heavy on my shoulders, the heat of flame at my heels. Behind me the fire chewed through the prairie with big teeth, taking out large chunks and leaving the rest, until long fingers of black with flickering flames at their tips spread out as far as I could see. The main fire was still acres away, a tumbling mass of brown smoke and distant orange glow. The rest of the crew was spread thin as butter at other corners, far from my position.

Nobody noticed the wind shift or the helicopter breaking from its pattern. The first indication I had that something was wrong was Jed's voice over the radio. "Get down to the west line, we need black down here."

His voice, usually calm, sounded tinged with worry. It took a lot to rattle Jed, and it sent a shock of alarm through me. Looking down the line, I could see it already; pushed by a sea breeze, the main fire had turned west to the place where there was no black line to stop it.

Steve vaulted onto the buggy, motioning me to follow. Roger, at the wheel, forced the engine to the top speed it could handle on this type of ground, about ten miles an hour, the big tires bouncing over the rocky fire trail. We had left Jen behind, but I knew that someone needed to stay to patrol our burning line.

On the buggy, Steve and I hung on for dear life. I knew if we fell off, we would be on our own. From the other side of the fire we heard Bill on the radio, saying that he was walking the dozer around to meet us. The helicopter buzzed overhead, en route to the work station to hook up a bucket for dropping water. Jed already knew it was more than we could stop with our tools.

We never lose our fires, I thought. Not the Fish and Wildlife crew. The Park Service guys did it all the time, with their massive burns; with fires that big, it was hard not to lose a few acres. But we never lost ours. Surely this would be just a little slop over, no harm done.

A month earlier, Jed had send us out to dig a fireline on the west side, since there was no handy jeep road there.

Suffering from hangovers and April heat, we had finished up late in the afternoon, tying the line knee deep in the black water of a cypress strand. Looking at the water we figured it was bombproof. Fire could not cross here. It was a sure thing. It was a good thing too, because the other side of our line was private property. Whose, we weren't sure, but it would be bad form to let our fire burn someone else's land.

Now when we reached the west line, the fire was pushing hard. Where we parked the buggy it had already flung itself across, a few spot fires blossoming in the dry grass.

"I'll stay here with the buggy," Roger yelled over the roar of an oncoming fire. There was no time to argue. He pulled a flapper from the top deck. As he beat out spot fires, more appeared. He was surrounded by knee-high flame.

Steve grabbed a drip torch as I stared at the main fire, hurtling toward us through the cabbage palms. It was a bright sea of orange, bearing down on us. It was the kind of sight that could make you freeze or run.

Steve handed me the torch. "Quick, we've got to burn this line out before it gets here." The intent was to rob the oncoming fire of fuel, stopping it in its tracks.

I glanced back at Roger, his lanky silhouette outlined by leaping flame, and prayed he would be all right. There was no time to do anything but run.

As I ran with the torch, sputtering flames from the wick out as I went, Steve at my heels, I could tell it was too late. The main fire was too close and our backfire would not suck it back into the unit like it was supposed to. Timing was everything with backfires and we had gotten it wrong. I dragged a line of fire behind me, but I could see that it was doing no good, just adding to the problem.

We stood for a minute, breaths caught. I was unable to believe what was happening to us.

"Run to the cypress," Steve yelled. We stumbled down the skinny fireline we had dug the month before, coughing in the smoke. I knew I should put down the torch, but I didn't, its flame flickering as I ran. There was no fear, only motion. This was what we had trained for so many times: finding an escape route to a safe place.

We stopped at the cypress, our bombproof line, but what had been wet before was now parched, the warm, dry weather of the last few weeks drawing the water underground. This would not stop the fire.

"Oh no," I whispered.

There was no other place to go; fire had cut off our route back to the buggy and safety. Soon it would pass through, aided by a strong wind, but for now we could not go back. There was no time to shelter up—besides, a fire shelter would be useless here in all this brush. We would be cooked inside. Heading deeper into the unit was a mystery: we could find cooler, damper ground, or we could be caught in head-high palmetto. I understood in a flash of certainty that we were trapped.

Steve pointed to the torch I still carried in my sweaty gloves. "Burn out a safety zone," he hollered, and I drew fire in a six-foot circle around us. It crackled through the dry grass and I stomped it out near my boots. We were using a slender circle of black as our last stand.

We stood for a moment in a ring of fire. The whole world narrowed down to that one moment. I peered through my goggles and bandanna. It felt like I was inside the fire, swallowed by it, although I was dimly aware that there was blue sky overhead, the whole oblivious world rolling by as we awaited our fate. Heat steamrolled over me and was gone, leaving a sticky aftermath.

The fire passed around us like a wave, almost touching at both sides, and was gone, headed for private land. Overhead the helicopter clattered with the bucket, dropping a silver stream ahead of us. I chattered as we headed back down the line, adrenaline chasing through my body.

"Can you believe it was dry? Why was it so dry? It wasn't dry last time we were here," I buzzed. Like a mosquito, Steve brushed me off. His face was inscrutable. Maybe he hadn't been scared at all. Quite possibly we had never been in any danger, even though it had felt like it. I decided not to ask.

We walked together, viewing what the fire had done. It had skipped over some places and burned others, with no real pattern to it. Here a clump of palmetto was burnt flat black; over there an entire swatch of muhlenbergia was untouched.

I stole a glance at Steve. It was easy to discount him sometimes; his round, boyish face made him look younger than he was, and he clowned around just like the rest of the crew. But as I had teetered toward panic, he had tapped into the icy core you needed to fight fire. He had saved us both.

The fire wasn't out yet, though the main front was long gone. There were clumps of grass still smoking, tendrils of flame inching up to the tops of the cabbage palms. We stopped at the intersection. I was afraid of what I would find, but Roger was still there, bracketed by leftover flames, his face streaked with ash. He wielded both the fire hose and a flapper, a one-man crew. He saw us emerge from the smoke and threw his hands in the air, grinning. "What were we thinking?" he yelped.

The helicopter reappeared, a shower of water dousing a hot spot. Bill clanked into the surrounding prairie with the dozer, turning over fresh earth.

"Better go get the spots," Steve said. Jen appeared by my side from farther down the line. We hoisted our flappers and headed out, chasing the fire through the tall grass.

We never talked about it again but I wondered. How close had we come to that dark edge? It was an ephemeral line I had visited before, coming close enough to fire's hot breath to feel the claw of panic deep in my belly. If others went there, they rarely mentioned it. "We pulled out our shelters as we hiked to the safety zone," they said, shrugging, or "We had to sit on a rocky outcrop while the fire burned toward us." The stories left out fear or regret or panic. I could only guess if the others imagined fire overtaking them, the shiver of sudden cold realization that everything had gone irrevocably wrong. It was our own secret, held close. After all, everything had turned out all right. We had made it. The old stories were just that, old bones we turned over once in a while in a training class. *That'll never happen to us. That will never be us.*

Fire Family

The rookie was just about to take a bite of a sour orange when Bill called us on the radio. "Bush hog broke down again," he said. "I need Roger and the toolbox."

Rescue calls in the swamp were common. Breakdowns were a fact of life on the limestone caprock, where the rock melted away from tannic acid, leaving deep, sometimes camouflaged holes. A buggy tire could drop in a hole, axles could break. A tire could flatten on a sharp piece of cut pepper tree. Worse still, a buggy could get stuck in the mud in a prairie, with no tree in sight to winch off from. By far the most infamous was Colding Prairie. This innocent-looking grassland about a mile from our work center and trailer claimed its share of victims. Several buggies became mired there over time, each necessitating a rescue by another buggy or even the dozer. On one long-remembered occasion, two buggies, one initially on the way to rescue the other, became stuck, with the drivers staring helplessly at each other.

Bill's breakdowns were legendary. Rattling over the rough trails with the dozer or the mower, invariably he lost a bolt or some other part. He then radioed in for help. Directions sounded comical to the outsider: "OK, you know where the shortcut is through Unit 8? Come through there, and then follow the main tram to the sweet orange tree. From there, go toward Tractor Camp then remember the place where we saw the big dead gator?" Finally the rescue squad, laden with spare parts, oil, and creepers, had arrived at Bill's location to begin the extrication process.

"Guess we better go save him," Steve said.

During our delay, Jen had been hunched in the grass, peering at flowers, her camo sun hat pulled low on her forehead. She swung back on the buggy and Steve made a three-point turn away from our destination, the trams. Going this way meant that we faced a deep, tire-sucking mud hole, but Steve didn't care. He maintained a poker face and swerved the wheels as we plowed through.

Mike hung over the roll bars. "Don't go to the left! Don't go to the left!"

Not one to question his own judgment, Steve steered to the left.

"Aw. You're gonna get stuck! You're gonna get stuck! And I'm sure as hell not walking back for the dozer!"

The buggy heaved back and forth, causing Jen and me to pale and clutch at the rails. We had heard stories of buggies going over, though we had never actually seen it happen. Unflappable as always, Steve increased speed and, mud-spattered and victorious, we sailed through to firmer ground, safe for the moment. We all relaxed.

"She's a burly babe from Michigan," Mike sang, making up a song about Jen and me, both from different parts of that state. Roger poked through the storage in the covered back seats, looking for wire to hold up an axle.

"Gah!" the rookie said. He threw the sour orange into the palms lining the fire trail. "You guys told me this was the sweet orange tree!"

We laughed. That trick never got old.

The refuge was a maze of intersecting trails established by the long-gone hunting club. Reminders of people were everywhere. We kept maps marked with important points: the tree fort, the hog pen, the clearcut. Hunters had left us with some hand-painted road signs, and we used them too, as mile markers: Cane Patch, Airplane Island, Lewis King Road. Our maps included big Xs for a forbidden fence line, where more than one buggy was sunk before it was declared off limits. They showed Alligator Alley, named after the interstate, a wet and swampy route where many baby alligators had been spotted. Often passengers on buggies scribbled on their maps, adding other details that might keep them from getting lost later.

I was half dazed and sleepy from the rolling of the buggy and the sun. The alligators dozing on the downed logs looked the same. As we passed them, they barely opened one eye. We paused by a small family, the young ones still sporting the stripes they would lose as they grew. Even though we were on a rescue mission, alligators were still worth a short stop.

Mike looked over the side with interest. Alone among us, he possessed a talent for calling alligators. A gulping noise made deep in the throat, something impossible for us northerners to duplicate, imitated the sound of a baby gator in distress. Make it long enough, and big alligators crept close enough to catch, which Mike did sometimes. If you held the gator right behind the neck, he explained, waving it in our faces, it couldn't bite you.

The rookie was from Oregon. All of his firefighting had taken place in the forest, not this strange and inhospitable swamp. His face, flushed from fear and sunburn, nearly matched the color of his ginger hair. He looked terrified, but we rolled on before Mike could do anything.

Steve was in a good mood because he was out of the office. He had been ordered to fill out forms today, but when he got a paper cut, that was the last straw; he had stomped out to the buggy and said he was coming with us. He sucked on a Swisher Sweet as he drove.

"I saw the hotshot woman of my dreams this summer," he said now, making a turn onto a trail marked with a street sign saying FRITZ BOULEVARD. "She had on one of those old-style faded Nomex shirts and was carrying a saw." He laughed, but he was half-serious, his face wistful. Steve was a little younger than I was, not quite thirty, but he felt the occasional loneliness all firefighters knew well. In any other town, he would have been highly sought after. Built like a brick house, all muscle with the added attraction of a sweet, bashful smile, Steve was just about perfect, but not here in South Florida, where the women valued other assets.

I nodded in sympathy. In town, thirty miles away, the people our age had jobs in banks and offices, if they even had to hold jobs at all. Many seemed to exist in a perpetual state of wealth and privilege. Coming into town after a fire, we stuck out. Our pickups barely ran, the mufflers too loud, oil leaving splotches on the pavement, and our hair straggled down our backs, untrimmed. We had farmer tans. We couldn't stay out late. One girl Steve took out complained about his beat-up truck. "I'll go out with you again if you get air conditioning," she said.

I told Steve about the millionaire I had met at a concert in the park. Tall and sandy-haired, at least fifty, he wanted to collect me like an interesting possession—a lithe, long-haired firefighter girlfriend, something to show off at parties. He liked seeing me in my Nomex and logger boots, but the novelty soon wore off.

"No man is going to wait three weeks for a woman," he said when I told him I was going to Texas on an assignment. I shrugged, packing my fire bag. The choice was an easy one, although I realized that I had been on the losing end years before with the smokejumper. It had seemed then that he had chosen fire over me. His back had been against the wall the way mine was now. It had been an impossible choice to ask him to make. You couldn't ask someone to give up the thing that made them sparkle, the very same thing that had drawn you to that person in the beginning. For the first time, I felt regret for how much I hadn't known.

Mike listened to our tales of woe. "Me? I'm just praying for a miracle," he said with a finality that discouraged future discussion. He crossed his arms and gazed into the distance, spotting Bill, a pacing figure in coveralls, smoking a cigarette. The bush hog, a tractor with a mowing unit attached, sat abandoned in a half-shaved patch of grass. "There's the fool now."

Mike was right to leave it up to fate, I thought, rather than waiting for someone to bail us out. If I never married, if I never found anyone who really got it, that would be all right. I had all this. I looked around the faces of my fire family, a deep affection snuffing out all the ways they could annoy me. I had Jed, who believed in me enough to keep trusting me with higher levels of responsibility: managing helicopters, directing the engine crew, running the aerial ignition machine, deciding the burn patterns on the fires we set and the ones we tried to put out. He let me teach fire classes to rookies, sitting in the back of the room with his arms crossed, signature floppy hat firmly on his head, a neutral expression on his face as he watched me try to explain relative humidity.

"It's the ratio of moisture in the air that exists right now to what it would be if the air was saturated."

Blank expressions.

"Okay, think of it as how much beer you have drunk versus how much you could drink before you passed out."

Dawning comprehension.

In the back row, Jed guffawed and gave me a thumbs-up.

Most of all I had this place, this intoxicating piece of swamp and prairie, pulsing with life and energy that matched my own. We didn't need to talk about other lives we didn't get to have, people who didn't accept the way we were. This was what my life was now, what my life would always be like, and I didn't need anyone else. This was, finally, enough.

Fire in the Blocks

We were slogging through another half-dark shift in the Blocks. Smoke drifted past, turning the buggy headlights into a dim smudge. The sky was the color of a bruise. The pump roared suddenly and stopped: out of water. Steve and Mike turned the buggy around and headed for the canal, their shoulders swaying in unison as the big tires rode unevenly on pavement.

This place we called the Blocks had a real name: South Golden Gate Estates. We called it the Blocks because it was laid out in chunks, straight roads paralleling each other. The Estates had once been touted as the largest subdivision in the country, prospective buyers flown over in airplanes during the dry season, no mention of the wet that would come and make living here impossible, turning chunks of it into murky swamp. Roads went to nowhere, canals were dug, street signs put up. Seduced by the low price and the promise of sunshine, hundreds of people were scammed. Bankrupt after agreeing to a court settlement requiring them to complete canals and roads in an area the size of New York City, the owners of the corporation bailed, leaving the Blocks to the squatters and the unsavory.

People dumped things they no longer wanted here: refrigerators, stoves, couches. We had heard of small planes landing on the stick-straight roads and of stolen cars brought here to be stripped. We even heard rumors that an American-backed Cuban army trained here in preparation for overthrowing their country. On a night like this when flickering flames cast deep shadows on the woods around us, it was easy to believe just about anything could be true.

It was eerie, driving down the abandoned roads, halting at the unnecessary stop signs, reading the street names, streets with no houses on them. It felt like we could be swallowed up in here, disappearing forever. It was not a place where you wanted to be alone.

Arsonists started fires here, guys we never saw but could imagine, scuttling down the roads close-flanked by overhanging trees, grass growing up through cracks in the pavement. They knew as much about fire as we did. Waiting until the wind was just right, they lit a match and backed away, unseen.

This fire started in early afternoon and took off right away, racing through the palmetto and cabbage palm like a train. It was a slow day back at the work center and we had given up all pretense of productivity, listening to the high-pitched radio chatter between the dozer guys and their dispatch. Finally we heard the words that mattered: Inaccessible, and deep swamp. Outgunned, the Division of Forestry put in an order for a swamp buggy and we were on our way. As we drove, we hooted and hollered, our good-natured rivalry over the parkies to the east flaring up again. We had beaten them to it, and we imagined them hunkered down in their engine bay, cursing us.

"Pound a liter," Steve ordered, meaning that we should guzzle one of our water bottles to prevent dehydration. Obediently, we all tipped our canteens.

Everybody wanted in on Block fires, mostly because they were big and mean and exhilarating. Even the city firefighters showed up with their wildland trucks, slowly patrolling the roads and directing high-powered streams of water at burning trees. Fire gnawed up the parcels like nobody's business, fueled by strong winds and land that had never been cleared. We tried to use the roads as barriers but they rarely held, sparks and burning palm fronds carrying fire to the next unit, the road suddenly a tunnel of flame. You could get trapped here, burned over in your buggy or your dozer. Without ever saying it, we knew that the terror was part of the allure.

The dozer guys who worked for the Division of Forestry loved it more than any of us. They tried to get as close as they could to the head of the fire, the place none of us were supposed to be. They just pulled their Nomex shrouds over their faces and set the blades down. They called it riding the curl.

We followed behind them in the buggy and on foot, lighting off unburnt fingers and putting out flames that crept across their line. Their lines were several feet wide, swaths through the swamp, cabbage palms lying skinned on the trail. It was like walking through a tornado's path: trees torn out by the roots, dirt thrown two blades wide. Dozers rumbled through the darkness as we walked with our headlamps, trying to avoid the falling trees they knocked down. The scars from their passing took a long time to heal.

Ahead we heard the back-up alarms of the dozers as they took another run at an obstacle. Trees fell like thunder; we couldn't get too close.

The rest of us waited, flappers in hand, until the buggy returned. The night was warm and humid and sweet-smelling despite the smoke.

The buggy lurched back to us. Steve jumped off when he reached a burning snag. He pulled on the hose reel and directed a straight stream at the burning palm. Water hit the fire with a sizzle. The trunk of the palm glowed red as if it were being lit from within. Steve circled it, adjusting the pressure. Pieces of burnt black chipped off, falling to the ground.

I took my flapper and smothered a small line of flames that crept across the line. I got close, close enough that I felt heat on my back.

I tugged my bandanna higher on my face. The fabric reeked of old smoke, a comforting smell that brought back memories of many other fires like this. I moved closer to the flames and sparks from a tree drifted down on me like snow.

"Your hair's on fire!" Mike said, coming up behind me. Smarter than I had been, he had dropped his Nomex shroud down around his face, but he had not fastened it with the Velcro closure so it flapped loose. "Flying Nun," I said, laughing at the two pieces of material that stuck straight out in the wind.

He dumped his water bottle down my back. We laughed some more. I inspected the ends of my braid, now a frazzled mess. Souvenirs of war, escapes in the nick of time: we all had them.

Steve clicked on his headlamp and the rest of us followed suit. The pale light illuminated our ash-covered faces. We all looked ten years older than we really were.

We gulped down water made stale from plastic canteens and gnawed on sawdust-tasting energy bars. The fire spread out before us in a glowing carpet. Fire flickered in the crowns of a few trees. Deep in the middle of the unit, trees flashed orange as pockets of fuel torched out. We looked, but they were no threat, surrounded by good black.

Humidity crept in as the evening wore on, making us shiver in our sweaty shirts. Roger stopped to shrug on his brush jacket, the stripes on the back and wrists catching the light from our headlamps. My pack chafed my shoulders and I hitched it up. Full of water and food, it hung low on my butt, bumping me as I walked. The hard edges of the fire shelter, contained in a lower pouch, jabbed ceaselessly as I moved. After these shifts I often found bruises, more battle scars.

The buggy ground slowly in low gear. Steve had commandeered an ATV and he rocketed around us to scout deeper in the block. Spooked, he quickly drove back out, eyes wide in his dirty face. "Found an abandoned trailer and a pot farm," he told us. "Come look." There was a smaller ATV trail that we suspected led to a bigger grow, but none of us wanted to risk it. There could be traps out there, anything.

"We have a reason to fail our drug test!" one of the city firefighters hollered, sniffing the air as he clung to the back of a passing truck.

Scorched plants in pots ringed the clearing. The fire started here, we speculated, an unwatched campfire making short work of the overgrown brush.

It was bad luck for the squatters, who could be watching us through binoculars right now, because they would have to start over in some other desolate patch of the Blocks, their cash crop destroyed. There were others like them out here, lawless and unrepentant, people either down on their luck or looking to make quick cash.

Jed had made me the Incident Commander on this fire, though the Forestry guys had their own IC who kept track of them. Jed liked to do that, give us each a chance. He would throw out nuggets of wisdom occasionally, seeing what we did with them. "Might want to think about putting two burners in there," he would say, followed by "or whatever melts your butter." He wanted us to think for ourselves, which was different from the way most people in charge did it. If we chose a different strategy, he would shrug, trusting us to pull it off.

This fire was by far the largest I had ever managed. It was winding up to be close to a hundred acres, maybe a hundred and fifty. Instead of fear, I felt only confidence. I had worked hard to get here.

"Fifty yards in," I told Mike and Jen, meaning that all smokes had to be put out that far in from the line, and they groaned but complied.

Overhead the first stars began to prick the sky. The lights of town, miles away, were a distant glow on the horizon. It was as if we were on some distant planet, one of fire and darkness and smoke. Roger was a lightning flash in the darkness, his coat caught in my headlamp beam. I paused, my chin propped on the handle of my flapper.

"Seen the dozers?" Roger yelled, waking me from my reverie. He stood on the fire trail, radio glued to his ear.

"Nope," I shouted back. We peered through the smoke. Cabbage palms, still smoldering, loomed out of the darkness, their trailing fronds turned malevolent. Small embers winked in the canopy like eyes.

Roger turned on his headlamp and shrugged. He was the dozer boss on this fire, tasked to serve as a line scout. His job was to head out ahead of the machines, marking the best place for them to go. For most of the day he had been flagging a line for two dozers to follow, dodging the falling trees they were uprooting. Being a dozer boss was always fraught with danger. You didn't want to get so far ahead of the machines that you couldn't see them, in case they had trouble getting through. You didn't want to get too close, either, because once the guys got on a roll, they didn't hold back. A person on the ground was hard to see and the tons of steel were hard to stop. Roger had been trying to stay ahead, but now he had lost his dozers. He hesitated, unsure of whether to go back the way he had come or to go forward.

I shrugged too. The dozer guys always did what they wanted. They rarely paid attention to us. They liked to see how close they could get to the fire's head, even though they would be toast if the dozer threw a track or got stuck on a high stump. They didn't really see the need for a dozer boss. They never anchored in to a bombproof point, instead doing what we called potato patching, punching parallel lines ahead of the fire like rows of crops. They had the utmost confidence in their ability. At the same time, I knew that they remembered a blow-up nine years ago, five years before I had met them, when one of their own was trapped in his dozer in this same area. Drought and a hard freeze, killing the live vegetation, had contributed to the unusual fire behavior that had prompted the firestorm. Marcos Miranda was only twenty-six when he died on his dozer.

A steady roar filled the air and two dozers suddenly appeared, heading in different directions. Their headlights swung wildly through the darkness. They bracketed Roger on the trail, ignoring his frenzied waves to slow down, and vanished.

Roger laughed. He plopped down next to me on a fallen log and took out his little yellow pad. "This fire makes sixty hours of overtime," he reported. We kept obsessive track, socking away each precious hour of time and a half pay in our skinny bank accounts. Sixty hours was nothing; we needed a lot more to make it through the year. Some jumpers made over eight hundred hours of overtime; this was small potatoes.

A cool dampness in the air signaled the rising humidity. Like magic the fire reacted, flames quieting to wisps of smoke. I could almost feel the division between day and night.

"Well, Darted Monkey. Better do something, even if it's wrong," Roger said, like he always did. I laughed like I always did when he said it. Roger hunted for his pulaski and found it where he left it, leaning against a half-burnt cabbage palm. He grinned, showing me the scorched handle. "Oops," he said.

Roger would get away with it, though. He always did. The rest of us were subject to chewing out for infractions such as this, but somehow Roger skated by unscathed. It should have made me resent him, but it didn't. It was impossible to be angry at Roger for too long.

Roger's hair was cut in the shape of a bowl. He was long and lanky enough that we sometimes called him Gumby. His cough annoyed me down to the last nerve. His nickname from hotshot days was Rummy for his beverage of choice. It was also Pedro for some unknown reason. One season, he broke his ankle playing volleyball and limped around with crutches on light duty, fixing broken-down vehicles and working on tools for months, but he endured this with grace. He didn't look like the hard-jawed, steel-eyed men I had come to associate with the good firefighters, the ones who had it together.

But Roger had it, that indefinable thing we all searched for. It was the ability to take one look at a fire from a helicopter or the ground and know instinctively what it would do, where it would go next, where it was safe and where there was danger. He was always one step ahead of us, his brain ticking through all the things that made up fire. It was impossible not to feel both jealous of and inspired by him.

We hiked back up the trail to join the rest of the crew. Cabbage palms sprawled across the fire trail, casualties of the dozers. Roger skirted the obstacles, still nervous after the ankle fiasco. He did not want to sit out another smokejumping summer.

On a typical day, Roger sat on the concrete floor of the pole barn, parts of the swamp buggy laid out around him. "It's like a puzzle," he explained, showing me. When he took off a brake shoe, he placed each subsequent part inside the one he had taken off right ahead of it. "You don't want to end up with an extra part after you've put everything together," he said, and laughed. He showed me how to drain the radiator, fix a head gasket, and replace a master cylinder. I handled each piece with reverence.

He talked me through loading the buggy on the trailer. Top-heavy and wide, it lurched onto the trailer, threatening to tip. I clutched the wheel, breathing hard. I pictured a tire sliding over the rail, a roll-over, death. Roger peered closely at every wheel revolution, motioning me forward. "A couple inches more. You got 'er," he reported, springing forward with the tie-downs.

He made me laugh too. When we gathered in the field office in the mornings to determine the day's plan, Roger hammed it up at the chalkboard. He squinted at notes that had been left for us by Steve, who had reluctantly headed to the office in town for a meeting. Interpreting Steve's cursory notes, Roger drew elaborate timelines. "Crew A goes to brush trams," he lectured, drawing an X on the map. "Meanwhile, Crew B drives to town to pick up supplies for the dump truck." More dotted lines on the map, and a few buildings to designate town. "Then once Crew B gets back, they'll meet up with Crew A and Crew A will pick up the dump truck and go up to Unit 15 to pick up trash."

Bill was always willing to throw a wrench into the plan. "What if Crew B doesn't get the parts?"

Roger hesitated, but rallied. He contemplated his map. "If Crew B can't get the parts, Crew A will snag Unit 5."

He put an X on Unit 5 and drew a stick figure for good measure.

We sat there for a while.

"We'd better do something, even if it's wrong," Roger concluded. The specter of Steve returning in a grumpy mood and finding us still sitting there propelled us into action. We jumped up, grabbing our tools.

Heading out to the road from the fire to catch up with everyone else, we hiked along in a comfortable silence. I thought about telling him how much he had helped me with his humor and patience. Sometimes I woke with my stomach in a knot thinking of all the ways I could screw up. Breaking an axle taking the buggy down one of the trails, not being able to start the chainsaw, making the wrong decision that would get someone killed. But if Roger was there I knew that everything would turn out all right. He gave me the confidence that I needed to make it through.

I knew that you just did not say things like that on a fire crew. We were supposed to be tough. If I did, Roger would make a joke of it; probably break out into a Barry Manilow song. Still, it was a dark night, the palm fronds silvery, a night when you could say just about anything.

But by now I had shoved my emotions down so far that they were hard to reach. "You're a hard nut to crack," a friend had told me once, and this made me equal parts elated and guilty. Surely this was not how I was supposed to be. Emotions were useless on the road; they made you do things you regretted, like thinking about a smokejumper who had forgotten me long ago. Paging a cop with steely blue eyes because even being with someone who didn't love you was better than another night of going it alone. It was better to make a clean break, to not let emotion drag you down. On the road you had to be aerodynamic. And on fires, too, unless you wanted to lose it and fall apart completely.

We broke out onto the pavement. The rest of the crew loitered by the pickups, drinking Gatorade. Roger hastily headed into the bushes with a roll of toilet paper and a shovel, much to everyone's amusement. I shelved the thought of talking to him. There was plenty of time left, years and years of seasons like this one.

I approached the dozers, parked on the road. Jen joined me, our boots scuffing the pavement. "Panther babes!" the dozer guys yelled. They leapt off their machines. They had started calling us "panther babes" and I kind of liked it. It was hard to feel like a woman, day after day sweating in twelve-inch leather boots, baggy green pants, and a dirty yellow shirt over an equally dingy T-shirt, hair scraped up in a ponytail and stuffed under a hard hat.

The two of us were called panther babes because of the Florida Panther wildlife refuge where we worked and because we were women in the fire world. The dozer guys thought that we looked good and could kick serious butt. They were not used to chicks with pulaskis; we were novelties. They had grown up here; their women were tamer.

"The panther babes are here!" the other guys echoed. They flocked toward us. The cute cop passed me an orange popsicle and winked. "Page me later," he said.

For a brief moment I was surrounded by men. It was always like this on a wildfire. I still stuck out when I took off my hard hat. In the mirror of the trailer bathroom, I often traced the new lines on my face, baked in from years of working in the sun. By now I had I noticed a few strands of gray in my dark hair. Sometimes I felt a vague undefined fear that I had let something wait too long, the marriage, the kids. But always, when I felt that way, I could lace up my boots, grown soft as slippers from years of sweat and ash, and head out the door. I could forget everything but what had to be done.

It was hard to think of giving this up. We all knew people who had. Suddenly they disappeared from the fireline. Some we would never see or hear of again; others we discussed in low, hushed tones as if they had died. Someone had finished nursing school and was grinding out shifts in a hospital. Someone else had gotten a real estate license. *How could they stand it*, we all did not say but meant, even though we knew nothing about their lives. It was a superior, condescending attitude, I knew, but it was also a justification. Otherwise I would think about friends left behind: Marla, my age, who had a real job and a baby. Robin, with a PhD. What did I have to show for my life? I didn't want to think about that.

Often we made plans to visit the ones who had gone off the line in the off seasons, but somehow those never really worked out. The man I found so fascinating in New Mexico had lost his shine in Tennessee; after quitting fire, he had become doughy with sloping shoulders and in fact had a girlfriend he planned to marry, a girlfriend he had conveniently forgotten to mention for the length of our assignment. The woman I shared a razor with at a fire camp shower turned into a party girl in town. A lot of times there was nothing to talk about; out of context and geography, it seemed forced. I sometimes thought that in our seasonal lives we were at our best, the most funny, the most interesting, the most alive.

Then there were those who hung in there into their fifties and sixties, moving slower and leaning on a shovel on the downhills. I watched them uneasily. I wasn't sure I wanted to end up like that either, stove up and shuffling like an old person in the mornings. It was easier just to think of myself in a suspended state, forever thirty years old.

When the fire was contained, circled in its entirety by dozer line, everyone took off but our refuge crew. This was what always happened. The dozer guys and the city firefighters who straight-streamed from the road got all the glory, all the newspaper shots. Because we had the buggy, we were the ones who could get into the interior to mop up the fire. We were unseen by the photographers and the news helicopters. That didn't matter to me. We watched the red taillights of the retreating city firefighters without comment.

"Bye, panther babes," the dozer guys called. "See you on the next one." They filed away to their dozers and loaded them onto the trailers. A final wave and grin and they were gone.

The refuge crew debriefed on the road under headlamps. This fire had no threat left. Even the flames crackling away in the interior would be gone by morning, victims of the cooler, damper air. "Good job," Jed said, swinging into his pickup for the drive back to town. I basked in the high praise.

A few more laps around the bumpy line, emptying another tank of water on the smoldering cabbage palms, and we were done. We would come back with a reduced crew to check it tomorrow, but chances were nothing would be left—perhaps some interior smokes slowly winking out. A black scar would be all that remained, but in this overheated, lush environment, it wouldn't take long to recolonize. First, small shoots of green grass would poke out of the damp earth. Next, the blackened palmetto would sport new green fronds and the cabbage palms would put out tentative growth. In six months it would be hard to tell that anything had happened here. Only the dozer tracks would remain.

The crew piled into the engine, the smell of sour Nomex making me roll down the window. Roger pulled out his notepad, recording overtime. The rest of the crew jabbered about how big the flames were, how thick the smoke. We headed north, our headlights piercing the night.

We swiveled in our seats, looking for the residual glow of our fire, a bright smear on the horizon. This was only the start, we predicted. This season, the fires would go big and blaze through the entire dry season and into the wet. Surely the next one would be the Big One, the granddaddy fire we all longed for but never quite found. Though none of us could describe what this meant, we kept hoping for it. No fire ever quite reached this status. There was always another one out there, lurking on the horizon, that would top them all. If we stuck with it long enough, we thought, we would get to it. We just needed to put in our time.

Sitting in the backseat, sticky with sweat, smoke clinging to my clothes, I knew that this was one of the hazards of firefighting. Like any addiction, you needed more and more of it. The rookies were satisfied with a little flame, but soon they needed more and more. Nothing could really satisfy us. We were drunk on fire, intoxicated and dizzy. It was a feeling we wanted to capture in our other lives, but nothing really came close.

"The best fire is the one you're on," we all said, but none of us quite believed it. There must be something even better out there, a fire with bigger flames, better scenery, more overtime. As we drove toward the work center I realized that I was living my life this way too. No set of mountains, no whitewater river, no person was ever enough. Each first kiss was going to be the best, each man the one I would never leave.

In the darkness of the pickup, I explored the tired faces of my friends. They sprawled over the seats, spent. They were my family, forged not of blood but of fire. We were bound together so tightly that I thought nothing could tear us apart.

Death at South Canyon

Some summers sizzled with the hushed breath of anticipation. In Idaho that summer, the summer that everything changed, stingy winter snow meant there was no melt and there had been no spring rain. Pine needles, cast from stressed trees, flattened under my feet in June. The berry crop failed early on and bears stormed hiker camps, clawing a woman's leg as she lay in her tent, kamikazeing into the trees for hung food bags.

I was back in Idaho as a wilderness ranger, but I was doing more firefighting than rangering. By the time July rolled around, it was common to be called out on a twenty-person crew or to smaller fires as a two- or five-person initial attack force.

This time we were somewhere in Colorado, or maybe Utah: the states had blurred together. I thought that it was the Fourth of July, or maybe somewhere early on the fifth. Time had lost meaning as it always did. There would be no fireworks for us, no burgers on the grill. Instead my twenty-person crew, a hodgepodge of forest workers culled from other jobs, had spent the last several days crawling through burnt sage, cold-trailing the edges of a massive range fire. Firefighters dropped in the heat; an ambulance visited daily, carting away the stricken. Our water was too hot to drink and we jockeyed for shade under a spindly juniper tree.

This fire assignment was remarkable only because we were led by a female crew boss, the first I had ever seen. Carol was slight, with a fan of blonde hair poking out from her hard hat, only five years older than I, but she kept the crew in line in a way I had never seen men do. Instead of leading out with a pace so fast we were running to keep up, hoping to break the weak and the unfit, she maintained a steady walk. Instead of barking out orders, she spoke in tones of gentle steel. Looking at us the first day, she had nipped any complaining in the bud.

"I know you'll be a great crew and not a bunch of whiners," she said. After that, nobody had the temerity to voice any dissent.

After the range fire, we waited, passing mind-numbing hours sweltering under a yellow tarp as camp folded around us. Everyone else was off to somewhere new but we were left to wait for our next assignment. Teams were reluctant to release crews once they had them; it was too hard to get them back, so here we sat. Somewhere in Moab an overhead team was deciding our fate. It could be days before they decided. Remembering Carol's admonishment, even the rookies were silent. We would move when we were told to move.

At night we lay in a row in our sleeping bags, watching lightning strike the Book Cliffs. "Money clouds," someone said. "Black forests make green wallets," someone else said. This was an attitude we could only carry off with other firefighters; it sounded uncaring when we said it, but we all knew fires deep in the forests were the only ones to joke about. The others, the ones that burned down houses, were off limits. Nobody really wished for those.

Finally Carol motioned for us to load up into our three Suburbans. We had an assignment, though little was known about it. The promise was enough to keep us wild-eyed at the windows as we drove a long way up a kidney-pummeling road that wound into the heart of the mountains. Obviously there was no threat to a town; the lights of the city were far below us. Rumors flew; apparently somebody important owned a summer cabin around here. We didn't care; the only thing that mattered was the fire.

A lone helitack crew member, the IC, met us and spouted off facts in high RPM: fifty acres, lightning strike fire, active on the south flank. "You're the only crew here," he said. "Some engines might show up, but we've had a big lightning bust on our zone. They just want this one wrapped up quick."

Only fifty acres? I felt the sting of disappointment, but there was no time to reflect on the puny size of this fire. It was a night shift again, no sleep for the second night in a row. In the dark we dug line, the feeble light from our headlamps inadequate in the deep night.

A warning passed down the row: "The IC said that there's a thousand-foot cliff somewhere out here! Look out for it and flag it!"

After the message had passed through a few people, it condensed down to "Big cliff! Heads up!" Little sparks glowed like eyes, a carpet of them as far as we could see. We fell into a rhythm of swing, dig, take a step. We sang all the fire songs we knew.

Digging line at night was different than during the day. Night shift carried with it a dim sense of unreality. Trees pressed in, smoke thickened, until I felt as if the boundaries of the world were narrowing to this one hour, this one moment. Night shift was deep and closed in, as if day would never come.

Somewhere in the night, Carol hollered to round us up. "Look, it's too dangerous to continue. There's a thousand-foot cliff out here and nobody knows where it is. Just hunker until daylight and we'll get started again then."

It was cold, the Southwest desert losing heat rapidly. We longed for our sleeping bags, safely back at the Suburbans, and a brave soul asked if we could hike back and get them. Carol shook her head. "If we go back and get them, we'll be taken off the line and not paid." Nobody wanted to lose the astounding holiday pay plus overtime that we were racking up. For many of us, the extra overtime hours meant the difference between poverty and survival. It was silly that if we encased ourselves in a single layer of down, we were considered off the clock, but we knew that the bean counters would see it that way.

Miserable and cold, I contorted my body around a flaming stump, waiting out the night. Sleep was impossible. One side of me was baked, the other chilled. To stay warm, I got up and poked around ineffectually with a pulaski. The rest of the crew hunkered down like rabbits under cover of the night.

At dawn a few shadows emerged from the trees. The crew member closest to me spit a stream of Copenhagen and pointed to an upheaval of earth. He belly laughed. "We've been digging line right next to a dozer track!" We trooped over to look. It was true; a fresh bulldozer cut wove parallel to our line. The whole shift had been wasted; we could have dropped back to this track and used it as our holding line. We could have burned out some fingers of the main fire to make a rock solid black. We could have tied in our line to the cliff.

Standing in the gray light with smudges under our eyes, we shrugged, too tired for anger. Why we hadn't known about this was a question nobody asked. We were too numb to care.

Giving up on this section of line, we moved on to a stretch that passed through a flat expanse dotted with scrawny juniper trees. Tawny grass, thick and hard to cut through, covered the plateau. Somewhere in my mind I registered the stark beauty of the shaggy-barked trees with a golden carpet underneath, but mostly I looked at the landscape in terms of how hard it would be to cut a three-foot line through it. We had been on the move for thirty-six hours now and the strain was showing. Diggers far back in line screamed angrily up to us in front. "You guys left us too much! Come back and get these roots!"

"We got the roots!" the man next to me snarled. "Do your job back there!"

Someone started to sniffle, dangerously close to crying. A rookie raised his voice. "I'm certified to use the saw. I took the class this year. Why won't you let me run it?" The designated sawyer, shouldering the saw, stubbornly refused to let it go. Others shrank deep inside themselves, not responding to jokes. They shuffled like zombies, repeating the same tasks: swing a tool, punch the dirt, pull it away.

I was somewhere in between. I reached the zone that skated between euphoria and collapse. I often found myself here; if I pushed through the initial blur that came from sleeplessness it was a place where cotton filled my ears, my boots floated above the ground, and I had gone beyond suffering. It was a mystical place that I had found only in running long distances and on the fireline. I liked this place. Nothing hurt here and nothing mattered.

In daylight the cliff was easy to find. It split the plateau, dropping in a crumbling line to the canyon floor, perhaps a thousand feet down. Though we could peer over and see smokes rising lazily below, the IC told us to tie our line in to the rocky cliffs and mop up the rest of the fire. It seemed risky to leave fire below us, but at this stage we did not question anything. We only did what we were told.

It was the worst kind of mopping up, the dry kind, a powdery dirt that held in heat that resisted our shovels. A couple of high clearance engines had shown up after all, on the same axle-busting road we had navigated, their crews wandering out to watch us but offering no assistance. "Can't help you out," said a portly engine captain, his shirt straining over his belly. "Hose won't stretch that far." The engine crews napped in the cabs of their trucks, booted feet sticking out the windows. I watched them with eyes that felt like coals burning, but my stare did not shame them into grabbing tools to help us. Why would they? It all paid the same.

In pairs, we poked and stirred at embers; we chopped and scraped at burning logs. Squads broke off and took coyote naps, little breaks to cut the brutality of sleeplessness, going on forty-eight hours now. When it was my squad's turn, I curled up under a tree with them.

The three men started talking. "Where do you think we'll go next? There was a lot of lightning a few days ago. Shit, I haven't been home in months. My dog doesn't know what I look like."

"Can I bum a cigarette?" the rookie asked. They puffed in silence. Sleep was elusive. If we fell asleep now, we reasoned, we might never wake up. At this point it was easier to stay awake and ride it out.

Finally there was no more mopping to be done and we reluctantly reconvened at the cliff. The longer I looked at it the more formidable it seemed, a crumbling massif studded with smoking trees. It was a dirty burn, small circular fires scattered through the forest with no continuity. Patches of unburnt forest circled the hot spots. It would be a tedious afternoon of crawling through the jackpots—piles of fallen trees—with the added spice of unknown fire below.

I knew without being told that this was one of the most dangerous scenarios there could be. Unseen, a spot fire could creep into the heavy timber and explode, running uphill toward us. We all glanced at each other but none of us spoke up. I thought of the unintended poetry of the eighteen watch-out situations we were trained to memorize and recognize: *In country not seen in daylight. Building fireline downhill with fire below.* I looked around again but nobody's eyes met mine.

I kept my mouth shut as we dropped into the canyon, using our tools to stop our slide. We were all used to being can-do; if we refused this task it could mean a black mark against our crew.

A new-booted member of our crew perched above us, radio in hand, acting as lookout. He was spared the hike down due to blisters but had to fight the urge to sleep. He was lucky, though, and he knew it. Sometimes blisters sent people packing to ignoble defeat, back to camp. He gave us an embarrassed wave as we filed past.

"Living the dream," the man next to me said without enthusiasm as we reached the first circle of smoke. It was slow, tedious work, and most of us didn't think anymore about the possibility that the smokes below us could kick up. Having dismissed it from my mind, I focused on the head-down work. Asses and elbows. There was only this time to get through, stir around in the ashes, take off the glove, feel for heat. Time slowed to a crawl. It seemed impossible that this would ever be over.

Mid-afternoon, a shout went up from Carol. "Line out, up to the ridge!" We weren't done, but slogged up anyway, used to following incomprehensible orders. On the cliff, we formed in a tight knot. Everyone shuffled at an uneven pace. I was incapable of conscious thought or decision. Sweat stuck my shirt like skin to my back. My hair lay flat to my scalp under the hard hat.

I scanned the rest of the crew. They looked as though they had been through a war, faces drained of color, eyes staring at nothing.

"There's a cold front coming in," the IC warned us, standing next to Carol. "Big winds on the way, a red flag warning."

This was the first time I had seen him in hours. His face was stretched tight with exhaustion, the freckles standing out in sharp relief against his pale skin. He looked about twelve, too young to be leading this fire.

"We're pulling you guys off the fire for the night because things could get extreme," he said in a tone that held no room for argument.

Some of us exchanged bitter glances. As tired as I was, as dangerous as I had thought the fire below the cliff was, I didn't want to quit. I wanted extreme fire conditions. That was what I was out here for, that heart-stopping rush of flame through treetops. Not a boring mop-up show. I didn't want to go home without a story. Even though none of us said anything, I could read the rest of the crew by how they dropped their tools in a pile and walked away. They felt the same way I did.

It was clear that we were beyond usefulness anyway, even though the rookie muttered about lost overtime. A few crewmembers hiked back to the Suburbans and delivered our sleeping bags. Nobody complained about losing overtime; the desire for sleep overruled everything. We lay in a line, all of us out in the open, old smoke filling my lungs. On schedule, the cold front arrived, the wind howling across the ridge, making sleep impossible without earplugs. I buried my face deep into my bag, smelling the dirt and smoke from a hundred other fires.

The next morning we staggered out of our bags like drunks. Someone broke out a hacky sack and a few of the crew got into it, executing wild leaps as they attempted to kick the little bag. Others hunkered on heels, waiting it out. The IC huddled with Carol. Sidling closer, I tried to eavesdrop. They looked serious, more than our fire warranted. Something was clearly wrong. I felt an unease start to spool through me.

"Gather up," Carol said. We straggled over to stand in an unsteady circle.

The rookies were bored, looking at their dirty fingernails. They didn't grasp that something bad was coming. The guys looked like they had crawled out from caves. Their beards wild, hair standing up, faces black, they barely looked human. I was sure I looked only marginally better. At least I had remembered to brush my teeth.

Everything on the plateau was black—rocks, trees, even the ground itself. Our faces, unwashed, were smeared dark. A spaghetti tangle of blackened hoses, attached to the remaining engines, stretched out across the landscape. Smoke filtered through the dead trees. The wind had blown all night but our fire had not grown any bigger. Below the cliff it still skulked around, sending up a myriad of thin gray smoke columns.

Some of us knew that 1994 was a dangerous fire season, one unlike any most of us had seen, except for the Yellowstone fires of 1988. Of all of us on the crew, only I and a handful of the others were old enough to have been there. This felt like that summer, a hush in the forests like a held breath. But even though some of us knew this, what I heard next was incomprehensible.

The IC stood in the middle of the circle. "There's been an accident," he said. He said something about a fire a few miles away, people killed on a mountain called Storm King, a fire named the South Canyon Fire. The cold front caused a firestorm and the fire roared up from below, catching them as they ran. "There were fourteen of them," he went on. Details were sketchy. That was all he could tell us.

"We're sending you guys back to your home unit," the IC said finally. We were told to pack up, that we were heading to Moab and then home. The rookie kicked at the ground. "We're not even done with our twenty-one days," he muttered. I glared in his direction and he shut up.

The ride down from the fire was silent, each of us occupied with our own thoughts. The deaths were starting to sink in, people just like us, somewhere on their own cliff. It could have been us, I thought, if we had not been pulled off the line. Even though our fire didn't kick up, it could have. All the signs were there. Why it didn't was a mystery to me. Water in the logs, patches of bare ground, luck, it could be anything. The rookie poked his head over the front seat.

"What's the matter with you guys? I heard this was the party Suburban."

I locked eyes with the others in the front seat. To the rookie, the news was incomprehensible and far away, people he didn't know. The deaths were not even real to him.

It was true that on the way to this fire we had laughed and joked, giddy with the promise of overtime and smoke. Everyone had wanted to ride with my squad. Now we were stunned into silence.

We were ushered to a motel instead of sleeping under our tarp, another sign that things had gone terribly wrong. Usually the overhead did not like to spend this kind of money on us. At dinner, rumors were passed along with the plates of food. "There were some smokejumpers who bought it," someone said. Smokejumpers. It was hard to fathom how this could happen to the best of the best.

A typewritten list went around, the names of the dead. I watched it pass from table to table. Most of this crew was green, and the names meant nothing. For many of them, this was their first fire. I watched each of them study the list. The names of the smokejumpers I knew snaked through my head, a kind of song. Mark, The General, Roger. Mark, The General, Roger. Were their names on the list? Three out of four hundred? What were the odds? Why was it taking so long? Finally I snatched it from one of the crew.

It can't be true. But it was, a name typed neatly in black and white. *Roger Roth, McCall Smokejumpers.* My friend Roger had died in a fire shelter on a mountain named Storm King.

Our Florida crew reunited in Roger's hometown, one hundred miles from where I had grown up, for the funeral. Six of us stayed in one motel room, all of us sprawled out on the double beds. How could we even sleep? Sleep was something we had forgotten how to do. Instead we told stories about Roger. We talked about how magical he was, his hands able to fix anything: pumps, engines, seven-layer cakes. How he optimistically fermented wine in white plastic barrels but sucked on the hose before it was ready, drinking up the profits. How he had just been there, part of the fabric of our lives, until he wasn't.

We paged through the number of times when we had gambled and won, the many times we were too far from our safety zones, the times fire outwitted us, making us scramble. I remembered our dive down the mountain on my first fire and standing in the circle of fire with Steve. The stars had always aligned. It had been so easy to believe that they always would.

It'll never happen to us.

We stood outside on a day that was so sun-washed and glorious that it broke my heart. I stole glances at the coffin, unable to really believe that whatever was left of Roger was in there. People far removed from fire clutched their programs, their faces uncomprehending. This was a violent intersection of fire and real life.

That would never happen to us. That'll never be us.

But it had. And it was.

Losing My Edge

The smoke was everywhere. It filtered in past the fragile skin of my bandanna into my lungs. It stopped up my nose. It lay in a thick net over my hair and my clothes. It blocked out the sun and turned the world into a cold fog. It was hard to remember a place without smoke.

It seemed like all the fires were in Idaho the summer Roger died, the Colorado blazes tapering off after he was trapped on Storm King Mountain. After being sent home early from the fire so close to his, I had thought about getting out. One woman I knew did. "I'm done with fire," she said, saying that the deaths had changed everything for her. Other firefighters declared that sticking with it was what Roger and the others would have wanted. It was hard to know, and so when a squad boss assignment came along on a fire near Boise, I grabbed my pack and got on the bus.

I could barely see Chester, a wiry Idaho farm kid, in front of me as we climbed the hill; behind me, Little Mike—nicknamed for his baby face and his age, barely nineteen—was hardly visible either. The hacking fire cough ran down the line as we struggled to draw air in. My breath stung as I tried to draw it in. *This can't be good for me*, I thought, but dismissed it.

"Inversion," Chester said. The smoke was holding down the sun until it finally gained enough strength to punch through. The inversion was actually a good thing. It meant that we could gain some ground on the fire. Once the lid lifted off the inversion, all hell would break loose, the ragged column of smoke straightening out, a plume in a hard blue sky. That was when everything changed and you just prayed that you'd made the right decisions.

Eleven was usually the magic hour. It would start with a faint breeze, barely enough to be noticed, just enough to sway the trees a little. Patches of blue started to appear overhead, drift smoke parting and closing like curtains. The temperature rose several degrees all at once and I began to shed the layers I had pulled on earlier that day. As I took long pulls of lukewarm water, the inversion shattered just like that. One minute, encased in smothering smoke; the next, bright sky.

The radio had been mostly silent up until now, mundane chatter about delivering meals to this spike camp or crews to that drop point. Once the inversion lifted it felt like a long exhalation of breath. The airwaves sizzled with sound and electricity. *Fire's picking up. Get on over to the division break. Fire's slopped over the line. I need an engine now. We're going to lose this thing.*

Now the helicopters could fly too; they droned overhead, the small bubble-windowed recon ships and the bulbous red sky cranes with their snorkels for sucking up water. Like giant insects, they dropped gracefully down to the river, filling up their 2,600-gallon tank in less than a minute. Tankers nosed their way like sharks through the sky, releasing retardant ahead of the fire in a fine mist of red rain. Crews were shuttled to places we had walked past that morning, lucky to have waited out the inversion down at camp. Looking down, I could see the yellow shirts of another crew as they labored to tie in with us. They had a long way to go, miles.

On our own ridge, I began to hear the fire. It sighed like a great pair of lungs. If I didn't know better I would think it was only the wind, far in the canyon below. Rows of drought-stressed trees marched away from me, down the ridge and into the gap between the crew and an unseen river. There were no trails, no sign that anyone had ever been here before us, as usual. The forest swept to the horizon as if it went on forever.

Deep in week two, it was easy to believe that nothing else existed but smoke and fire. That there would always just be fire camp, a motley set of yellow tarps angled over poles, the same lukewarm eggs in the mess tent, the row of school buses parked in the dying grass. The community shower trailer, steamy bodies of women gathered around inadequate spray, the paper towels the camp provided sticking in patches to my damp skin. My wet hair, still smelling of smoke, slapping against the back of my yellow shirt, freezing in the high elevation summer chill as it dried. The row of blue porta-potties, reeking of thick blue chemical or, worse, overflowing with sewage. The march up the hill in the inversion, the smooth handle of the pulaski sliding in a gloved hand.

Somewhere in the back of my mind I knew that normal people were on vacations. That the cute cop I still thought about sometimes, slippery and elusive as a fish, was back in Florida waiting for me to return. Or not. Perhaps he had hooked up with another long-haired girl, one who did not go away. I couldn't think about that now.

With the lifting of the inversion, we were on the move. We double-timed it, making up ground. Saws whined as the crew moved downhill, single file, through a mass of bone-dry manzanita. The fire was somewhere below us. I could hear it snaking around, a hesitant snap-crackle-pop, but could not see it.

When we stopped for a water break, I looked over at Joe, one of the other squad bosses. Dave, the third squad boss, was farther along the line with his men. Joe rubbed his fledgling fire beard and wore a look of habitual concern. I wondered if he was thinking the same thing I was. There was a line we always walked, a tightrope between caution and risk, and sometimes it was hard to find my footing. Some fell too far on one side and never accomplished anything, cutting line miles from the fire, pulling crews out too early. Others rolled the dice the other way, taking chances, driving too fast, dropping into a canyon too soon. Most times they made it out alive. Finding the balance seemed harder and harder these days.

Although this fire blended in with every fire I had ever fought, this time was different. Barely a month had passed since Roger had died, and I wondered if I'd lost my edge, become too hesitant. There was no room for this on the fireline, but every step I took seemed fraught with danger. Each wisp of smoke could be the one that boiled up the canyon to trap us, each tree the one that fell and took someone out.

Of course I had thought of those things before; we all had. Steeped in tales of fires gone wrong, it was hard not to measure each step. This was different, though. I felt I was walking through a new and treacherous land. The weight of a potential mistake settled heavily on my shoulders. Straightening my stiff back, I watched the four people assigned to my squad. They were bent over their tools, oblivious, trusting that I would keep them safe. I was no longer sure I could do that.

I looked around for our crew boss but didn't see him. He was prone to solitary hikes up the line, neglecting to tell us where he was going or when he would be back. I only knew him by reputation. He was old school, the strong and silent type, not inclined to joke around with us. He threatened to send us home for minor infractions. He hung on to the map and the shift plan, refusing to relinquish any information. Drunk on power, he swaggered off alone, returning hours later without explanation. "Where's the main fire? What's it doing?" my squad members asked me. Without information, I mumbled an answer, not really sure of anything. Dan and Joe were in the same boat, growing more visibly irritated with each passing day.

Named after the gulch where it began, this was a stubborn fire that blew up every afternoon, as if it were on a timer. I could set my watch by the milestones that punctuated the days: dawn, the trainee crew boss screamed for us to get up, an unwelcome task delegated to him by the crew boss, who had slipped away to the mess tent. Eleven, inversion shattered. Two in the afternoon, fire broke free. We could be pulled off the line to a safety zone or we could keep, chasing it. There was no clocking out around here, no automatic lunch breaks. Instead we pawed a sandwich when opportunity allowed, chewing as we worked. We snatched catnaps when we could, leaning our heads against rocks. We learned to be opportunistic and sneaky.

We had been digging line all day, falling into a rhythm, something that we could keep up for hours. Our conversation flowed and ebbed as we worked, an endless river. Sometimes one of us would pick up the thread of a conversation begun and abandoned hours before as if no time had passed.

"On day sixteen, we'll try my apricot face scrub," a rookie named Sue promised Deb and me between pulaski swings. During hot, dirty, tiring work like this, it was good to have girl things to look forward to. Most of the time, encased in layers of grime and surrounded by men who belched and farted with abandon, I didn't feel like a woman anyway. I suspected I didn't look much like a woman either. I was somewhere between human and animal.

Hygiene was the first to go, some of us not even bothering to shower. The guys especially would forgo the shower tent, announcing that they planned to wear the same pair of crusty underwear for the duration. They might rub a bit of toothpaste on their teeth but that was as good as it got. They slept with boots on. They refused to change their shirts, which became so stiff with dried sweat and oil that they could stand up by themselves.

Food I would not normally eat was shoveled down in vast quantities. The fireline diet was heavy on refined carbohydrates and sugar. I ate everything and wanted more. Small brightly colored orbs of fat and sugar made up our main diet, Skittles and M&Ms, along with peanut bars for those who could stomach them, supplemented by MREs if we were in spike, or by mass-produced, bland meals in camp. We lined up with our trays, hoping for the best, usually disappointed. Sue, a vegan, was out of luck and had to resort to swiping the single-serving peanut butter from breakfast to spread on stale toast. She whittled herself down further to bone as the weeks went on.

Some of us started out earnestly planning to line up for the pay phones at camp, but this resolve soon faded with the wait time. It was not only that, I admitted to myself—life outside of fire camp became less real as the days passed. It was easier to forget anyone left at home and pretend there was only this and would only be this.

Because I would be out here for three weeks, I had to buckle down and take it one day at a time, the way you always had to on a fire. If I started thinking about twenty-one days, twenty-one days of swinging a pulaski onto rubber roots and through thick duff and sparking off rocks, sending a shiver of pain up my forearm and into my shoulders . . . if I thought about that, I was done for. The same with thinking about twenty-one days breathing into the smoky bandanna around my mouth, twenty-one days of restless sleep under a tarp, twenty-one days of hiking in single file up the same damn mountain. I had to break it down into hours. That was the only way to get through it.

But even as it was something to get through, it was something to be savored. On the fireline, my life was fluid music. It was passion and intensity in doses I had never found in real life. A month after Storm King, I was still deep in this life, but I was beginning to think of tunneling my way out. The only problem was that I was not sure there was this kind of brightness to what I thought of as an ordinary life. What would I do to replace fire? Who would I be without fire to lean against? I had no answers, so for now I fought fire.

A Minnesota crew worked ahead of us, gamely scratching a line with more enthusiasm than skill. We leapfrogged them, pushing past their ragged front, but all of us, even the rookies, knew that we would have to come back and improve their lines. They had been pressed into duty due to the severity of the season, kicked out on the line of their first wildfire after a week of training. Because we needed something to complain about, they were the target. For one thing, because they weren't in shape, they got helicopter rides while we hiked in, an hour and a half through dense smoke. "There go the Mini-Scrotums," Little Mike said meanly as their line huffed up the next rise.

Deb and I had pulaskis and worked at the front of the line. "First Pulaski" was the coveted place to be, only gained after you had proven yourself capable. It had taken many fires, maybe twenty, maybe more, for me to move up to this position. We were the first women we knew who were designated first pulaskis. It was basically unheard of in our circles, and we knew our status was tenuous. As first pulaskis, we set the pace for the whole crew. Too fast, and there were long gaps of fuel where fire could sneak across. "Take more!" the followers would yell. Too slow, and the crew bunched up, waiting for room to swing. "Bump up!" they screamed. They never had to do either with us. Our pace was just right.

Deb and I had been roommates for the past couple of summer seasons, and we had squared off on either end of a crosscut saw, cutting out deadfall that littered the Idaho wilderness trails. Stronger than any other woman I had ever met, she always had a smile on her face. She and I were the perfect partners on the fireline.

"I wouldn't be caught dead with a shovel," Deb said, as usual mirroring my thoughts. "Or be the drag queen," I agreed, pointing out the last unfortunate who had been saddled with the rake. These poor souls had to trail along, last in line, forced to deal with all the problems the others had left. Too thin of a line. Not enough depth. Roots, fat and bulbous, ignored by the slackers ahead. The shovel and rake bearers were always in bad moods.

There were always men pushing the pace. They wanted to be first and they let us know it. By seniority and the grace of the crew boss, Deb and I stubbornly held our place. We would give it up for nobody.

As far as fires went, this one had been sweet so far. Our camp was next to a ranger station in the cottonwoods, a small and quiet place devoid of the usual generators and bright lights that characterized most fire camps. We hadn't had to pull any night shifts—another luckless crew was stuck with that, emerging bleary-eyed from the woods as we hiked up the ridge in the morning. The only thing troubling me was the ache in the back of my throat, caused not by any sickness but by memory.

Right after Roger died, it seemed like the thing to do, to keep fighting fire, to continue to love the thing that claimed him. He would want me to do this, I told myself. And I was angry, seized with a desire for retribution. *I'll go back on the line for him*, I thought, to prove his death wasn't in vain. Now on this mountain it seemed futile, a useless tribute. I wanted to be angry at fire, to renounce it like others had. At the same time I was still drawn to it. Though I didn't show it, I was torn apart by grief and uncertainty. I was looking for a sign.

Sue had committed the rookie sin of too-small boots, and her toe throbbed from hitting thick leather as we hiked, but she was hanging in there, her chin-length blonde hair swinging as she dug. We sang Johnny Cash songs, the ones you always heard on fires, about walking the line and rings of fire. Because we were going indirect, not close to the main fire, it was easy to be complacent, in spite of everything I now knew.

That'll never be us.

During a lull, Chester tried to teach me to spit properly. "I can't believe you've been around this long without knowing how to do this," he teased me, hawking a big loogie onto the dry soil.

Because things were slow, we stopped for lunch, pulling out MREs. Fierce trading ensued as people tried to snap up the edible portions. "Awesome! Bean burrito!" a lucky soul exclaimed. My worries dissipated for a moment and I leaned back against my fire shelter, taking it all in. This was all so sweetly familiar. For a minute I could forget.

We were sitting on a mountain above a wildfire, but it was easy to forget this fact. It was hot and sleepy there on the shoulder of a sandy mountain, bees droning in the head-high manzanita. Small brown birds flitted among the pines. The sky was cloudless and benign. There were no other crews in sight and the helicopters had fallen silent. I had had little sleep and I was lulled into a meditative state of complacency.

Despite Roger, despite everything, when the fire blew up, it caught me off guard.

"It's slopping over!" someone screamed from above us. "Bump back to the safety zone! RTO!"

RTO, reverse tool order. When someone said this, especially in a voice tinged with panic, it meant we were to about-face and get the hell out of there.

The drag queen was now in front and Deb and I brought up the rear. Breathless, in a long, snaking row, we hurried down the hill, abandoning our line. As we headed cross country through the green, some people fell and got up quickly. There was an undercurrent of panic, but nobody wanted to look like they were scared. Nobody abandoned packs or tools. Nobody asked why we didn't have a lookout, like the fire rules said we should. Nobody brought up the fact that our safety zone was a mile away, too far for comfort. Nobody brought up South Canyon at all.

This could be it, I thought. I mentally reviewed the procedures for deploying a fire shelter. Shake it out of the bag, making sure the wind did not snatch it from your gloved hands. Step in between the straps attached to the bottom. Fall to the ground and puff it up. Wait. Breathe.

But there was no open place to pitch one here, just a tangle of brush, no safety zone, and we ran on. Was this what Roger was thinking on Storm King in the one moment when he knew he could not outrun the fire? I imagined the hot gases filling my lungs, closing them up in a silent and invisible struggle. I imagined the flames overtaking us, burning through our clothes. But unlike him and the thirteen others, the nineteen of us made it. At the bottom, safe on a dirt road, I paused and looked up.

A huge smoke cloud billowed thousands of feet in the air. As I watched, the fire gobbled up the dry forest, no doubt jumping our line.

The crew boss showed up from somewhere unknown, his face impassive. I wanted to ask him what happened, how we got it so wrong, but I didn't dare. A rattling of tools heralded the Minnesota crew, wide-eyed and chattering a mile a minute. Their dirty faces were stretched in wide grins. Danger! Big flames! This was why they signed up. Watching them as we waited for our bus, I was reminded of my first fire and the day we escaped a fire much like this one. It seemed like a long time ago. Though I was only thirty, I felt ancient.

And why was this still happening, I wondered as I slumped into a seat. Hadn't we learned anything from all these reckless runs to safety? Why did we keep doing the same thing, over and over?

I pressed my sweaty face against the glass. Around me rookies hooted and hollered, pointing out places where the fire was running high in the canopy. A month ago Roger had died in a place like that, where fire and drought-stressed trees and good-intentioned policy had collided. In all of our classes, the rule was always that if we had to run, we had gone too close to the edge. At the same time, an unspoken weight lay on our shoulders: get in, put it out, do what it takes. It felt like we pushed that line on a regular basis.

The rapid downhill run pushed Sue's ailing toe over the edge. Unable to walk, she was sent home. Deb and I mourned the loss of the apricot face scrub, which Sue thoughtlessly took with her on her freedom ride to civilization. Now there was nothing to do but wait it out until Day 21.

The fire blew past our lines each day. The camp swelled with newcomers: more crews, jumpers, hotshots. The overhead team scurried through camp with worried expressions.

We sat in camp in the evenings, debating our fate. Our crew boss was tight-lipped, keeping information to himself despite our pleas.

The fire was big enough to be divided into divisions, with different crews responsible for each one. We were inexplicably sent up to Division D, which we were convinced was completely out, to grid for smokes. The hotshots grabbed the plum job of building hot fireline up where we were the day before.

Gridding was its special form of torture. Gridding day after day could break you. Joe grumbled loudly as he tried to force people into line. "Spread out in formation," he screamed, near the end of his rope. We were supposed to walk parallel to each other, scanning the ground for wisps of smoke. If one was seen, the two crewmembers closest to it stopped and dug it out, stirring in cool dirt and spraying water from piss pumps in thin bursts from a long nozzle. The rest of the crew just chilled, waiting. When it was done well, it was a breathtaking sight: a solid line of yellow shirts and green pants, moving in unison over a baked hillside.

But this crew balked. Some freelancers raced ahead, wanting to be done with the grid, pointedly ignoring smokes nearby. Our crew slacker halted in front of a tangle of bushes and flagrantly moved around them without searching. Deb and I muttered under our breaths as we crawled through the spiky branches, our shirts acquiring slashes of black.

Joe and I looked at each other again. We knew that a crew that did not gel together was a crew that was in danger of dying separately. We had seen examples all through our firefighting careers: the woman separated from her crew, dying in a fire shelter along a road, everyone else fleeing without her. The man mistakenly left behind as two engines bolted for safety, each thinking he was aboard the other one. This crew ignored our warnings, though. They didn't want to hear about Storm King. They believed that it wouldn't ever happen to them. They were smarter, faster, stronger. I used to feel that way too. But I knew now that it could. Every move I took on this fire was tinged with that realization.

As usual, the crew boss was nowhere to be found. We clung to one slope like rock climbers. "We're gridding in the green!" Joe exploded. There was obviously no fire here, but someone up in Operations had decided to hang onto us, so we had been sent to hide on a dead section. If another fire got hold of us, we might never come back and so we were rat-holed, hidden away.

"This is going to be a twenty-one dayer," I predicted. "Twenty-one days on the same fire."

Though we sometimes spent our whole assignment on one fire, often we got moved from one to another. I had noticed that more and more fires were becoming what were called project fires. They lasted months sometimes. It wasn't unusual now for a crew to spend twenty-one days on a fire, go home for their two days of rest, and come right back to the same place.

"We dug the first line when it was five acres. We'll dig the last line too," Joe boasted.

We hiked back to fire camp every night. The camp women, the ones who filled out our timesheets, sat at the mirrors in the shower trailer and blew their hair bone dry. One carefully laid out a full complement of makeup and methodically lined her lips as we trooped past in our disposable towels, clutching our heavy fire boots. Outside, I rolled my eyes. "Camp slugs," I said. Deb snickered. In spite of my aching arms, I said that I never wanted to be them. Most of the time, this was the truth.

The boys seemed to like them, though. They leaned out of the windows of the buses that transported us to our drop points, dangerously close to falling out. "Look at them Wranglers!" Chester yelled, earning a reprimand from the crew boss. "I wish there were more women here," Little Mike said, and then laughed when I reminded him that there were women on the crew. "I mean, real women," he elaborated, digging himself in deeper.

Rumors flew fast and furious as we forced in the calories. The mess tent was almost empty; the hotshots had clearly been sent to someplace more exciting. Someone returning from the medical tent reported that a demob list had been posted, the death knell for this fire. I haunted the information board, where newspaper headlines blared of the West on fire, houses burning, and states of emergency declared. The rookies high-fived each other. "We'll be gone all summer!" they whooped.

Some of the old-timers sat in the mess tent and talked about South Canyon. It meant nothing to the rookies and they ran out after eating, intent on showers. I lingered near the gray-bearded men who wore somber expressions. One of them said that Roger died because he volunteered to carry a cubie down to people working below. It wasn't his job but he did it anyway. If he had stayed up above, he would still be alive. I also heard that he died next to another firefighter who had crawled inside his shelter, perhaps running out of time to open his own. When the other man crawled in, flames came in with him, compromising the shelter. Sharing a fire shelter was something we had been taught only to do as a last resort. The images haunted me, wearing a groove in my brain. I was aware more than ever before of the danger we flirted with as we walked the line. Sometimes I was paralyzed by what lurked below.

After a few more unproductive shifts on Division D, we reached the end of our twenty-one days. We had narrowly missed being assigned to rehab, generally a dirty job of reseeding lines and breaking down berms without the interest of flames. The Minnesota crew, shouldering rakes, stared at us with ill-concealed envy. Arriving after us with a few days to go, now they were trapped without even the hope of hazard pay. "Sucks to be you!" Little Mike chanted to their receding backs and was reprimanded by the crew boss.

We signed off on papers proclaiming that we had turned in the lanterns and tarps that we borrowed from Supply and boarded our school bus for the last time. Among the rookies, the mood was glum. For some of them, this could be the end of the road. Once they appeared at the home unit, their supervisors might decide that the trail work was more important than their desire to go back out.

I stared out the window as the bus fired up to take us home. From there we would disperse back to our bunkhouses and regular jobs. Though I didn't say so, I hoped that this was the last fire of the season. This one, coming so soon on the heels of Roger's death, had been too raw. He was with me at every step I had taken on this fire. At night in the tent I clutched his worn ball cap close. Every day I fought back the sting of unfamiliar tears.

The unwritten rule was that you checked your emotions at the door when you signed on to fight fire. There were good reasons for this. If you fell apart on the line you could die, taking others with you. You could make bad decisions when swayed by emotion. I swallowed it all back as I sat on the bus, but I could feel it, a stone caught in my throat. I remembered the girl crying on the Wyoming mountain years before, and how we had ignored her, unwilling to address her pain. Now I felt sorry for what we had done. I wasn't sure how much longer I could keep my own tears down.

The bus lumbered out of fire camp and wheezed up the incline toward Stanley. As we topped out on Banner Summit, the rookies bounced in their seats, pointing out all the smoke columns from all the fires they could see. They speculated wildly on our chances of getting assigned to one of them. It didn't matter that every fire was basically the same: long, sweaty days of digging in the dirt. It was easy to forget that part and they did, already making up their own fire stories about this one. The flames got larger, the danger more intense. They leaned forward, willing the bus to go faster so they could get on with the rest of their lives.

I sat back. I wasn't sure what the rest of my life should be. When I talked about leaving fire, the others looked at me with a mixture of horror and pity. Maybe I was too old, I could hear them thinking. *She doesn't have it anymore.*

Maybe she should get a camp job. Wear white sneakers instead of fire boots. Bring newspapers around to the spike camps.

"I couldn't ever take a desk job," one of them told me earnestly. She was twenty-three, with no idea of how easy it was for her. The new girls hiked up the hills and pulled hose like it was their birthright. A woman on a crew was no longer remarkable. It was the way it should be, but it stung. Didn't they know how hard we worked for them to be able to do the very things they were now casually doing? They had no idea of what we had endured, the taunts, the jokes, the endless eyes on us as we had walked the line. And not just us, but the pioneers before us, the rare women whose footsteps my generation had followed.

Winter was coming up fast. Steve called with tales of the swamp as dry as it had ever been, hurricane-toppled trees making jackpots that would burn hotter than we had ever seen. There were fires already, he said. I pictured the flames blossoming up like orange flowers in the Blocks, chewing through palmetto and cypress and sawgrass with an indifference that mocked us all. Everyone was coming back for the season, Steve said. Even Jen. Everyone except one.

After Roger

The first winter season without Roger was the hardest. I stood knee deep in a trout stream in Montana, stalled on my cross-country drive. I watched the tip of my fly rod bend and my orange line float through the sky and dip beneath the water. On the river there were other fishermen, but we did not speak. Sunk in our own sorrows, we cast and cast again. It felt like we were all searching for something that we would never find.

I almost didn't go back. In the sage-studded valley where I had spent the summer there was a cabin I could caretake, one of my friends had said. I would have to fix fence, but he knew I was capable with a wire stretcher and a post pounder. There was no running water, but there was a hot spring down the road. Temperatures plummeted to minus thirty often, and snow piled up to the rooftop, but these mountains in winter were a magical place. Why did I want to go fight fire in a swamp? What was so great about that? Why not stay?

I thought about it. The rough gray mountains danced in a jagged line along the horizon. Snow clouds flirted with the granite faces of the highest peaks, dipping over them, then retreating, never committing to full cloud cover or sun. The Sawtooth Valley sprawled below, a big bowl filled with sagebrush. Below my feet the river moved busily out of sight. It was so beautiful that it made me ache to leave. Not for the first time, I wondered if I would ever be able to love anyone this way. My heart was wide open, the way I never let it be for a person.

Already the aspens up Fishhook Creek had dropped their yellow leaves, as big as plates. Snow frosted the passes, sending the outfitters down from the high country early. The sun hung low and sullen in the sky, barely making it into our valley before it set again. It would snow down low any day. Unless you wanted to be stuck here, any fool would know it was time to go.

And there was this: not going back, the year after the fire, felt like giving up on Roger. Maybe back there, where we had lived together, I would be able to put an end to the dissenting thoughts had run through my head on the last fire. I could figure out how to bridge the gap between loving fire and fearing it. I could figure out the rest of my life.

Back at the Florida Panther work center, it seemed like we all worked at trying to forget, but it was impossible. There were signs of him everywhere. His charred pulaski from the long-ago Sixty-Hour Fire hung on the wall, an unfortunate precursor to tragedy. The swamp buggies that no longer ran so smoothly without his magic touch. His locker, now used by a new white-bearded guy named Jim. When Bill broke down, we fumbled with unfamiliar tools, wondering what Roger would have done.

Jen soldiered through the season, but I doubted she would be on the fireline for long. An impenetrable sadness surrounded her, one I could not break through because I didn't have the right words. What did you say to someone whose fiancé had died in a firestorm? "The memories we had, things the two of us did, now I only have them," she said once. And: "I was lying in a tent eating popcorn when he died. *Eating popcorn.*" She was strong, though, stronger than I had ever realized. She shrugged on her Nomex and carried her torch with the rest of us, showing no sign of weakness. If she cried, she did it in her room at night when none of us could hear.

It was a grim season, punctuated with brief moments of laughter. Jim was deathly afraid of snakes and frequently leapt like a ballerina through the muhly grass as we lit flanking fire. One afternoon we were paralleling each other when I saw him fling his drip torch and run.

"What?" I said, impatient, as he gained my side by the road, eyes wide in fear.

"Huge snake, the biggest I've ever seen, I almost stepped on it," he babbled. "So I threw the torch at it." I couldn't hold back laughter. The main fire was too close and we were forced to wait for it to blow through before retrieving the torch. Roger would have loved this. I could almost hear his raucous laughter as Jim and I stood there.

In a moment of perfect timing, Jed came down the fire trail on an ATV and spotted us guiltily watching the main fire approach.

"What's going on?"

"A snake," we babbled. "Our torch!"

Jed was not one to mince words. "I have to get back to the orifice," he said, his name for his cubicle. "You'd better go in there and get the torch."

We stared at him. "But the snake—" Jim ventured.

Jed stared back. "Better get that torch," he repeated, firing up the ATV.

His face blanched white, Jim tiptoed through the burnt grass to retrieve the torch, jumping at every rustle of grass. I laughed at his timid approach, the sound unfamiliar and good.

He waved me over. "It's still here!"

An eastern diamondback rattlesnake as thick as my forearm lay curled against the drip torch, eyeing us balefully. A distinctive rattle filled the air.

I looked over my shoulder. Jed had disappeared in the distance. By mutual consent, Jim and I bolted back to the fire trail. We would wait here, I told him, as long as it took.

Another day we were burning Unit 2, lighting the same prairies that Roger and Jen and I used to burn. Swaying on oversized tires, the buggy returned from its mission of spraying down the same billboards out by the highway that we always sprayed, to the same unit we had burned three years before. Sometimes it seemed we were caught in an endless loop.

"I'm praying for a miracle," Mike said. He said this a lot, and none of us knew what he meant by it, but we laughed each time anyway. We laughed this time too.

"Hop on, Darted Monkey. We need to go draft," he said, meaning that we had to go pull water from the pond to the buggy's holding tank. I grabbed onto the rail and pulled myself up to the front deck. The sheet metal under me was hot from the fire the buggy had passed through. Everyone else climbed on too, finding places to sit wedged in by the coolers or on the back seats behind the rails.

Fakahatchee Mike, on loan from the state park of that name, hunched at the wheel, chuckling over my nickname. Our Mike explained, waving a cigar expansively: I was famous for refusing to miss our daily run. Even a creaky knee did not stop me as Mike and I jogged along the highway, breathing in the fumes from the traffic between Immokalee and Everglades City. We used to be allowed to run on the dirt fire trails, but our new refuge manager laid down the law. "If any of you gets attacked by a panther, that would set our program way back," he said. Often the interns had pointed out where panthers had crouched just feet away from us, watching us run by.

Enough rusty vans passed at a lingering pace, leering faces peering out of windows, that most of us would have rather taken our chances with the panthers. But as seasonals, we had no job security. Performance ratings were our only ticket to another job. We had to run, and run fast. So here we were on the highway, our joints taking a beating once again.

I limped along, grimly determined to finish the run and to beat Mike while I was at it. Mike, never known for mincing words either, started to guffaw as he kept pace.

"What?" I snapped. "What are you snickering about?"

"You! The way you're running! You look like a darted monkey!" This struck us both as so funny that we laughed all the way back to the work center, forgetting our rivalry. The nickname had stuck.

Laughter was what Roger would have wanted. He used to try to elicit chuckles from the grimmest souls: Steve, when he was mired in paperwork; Bill, when his back was aching. Laughing would not bring Roger back, but it could help the ones he had left behind.

We cut across the black grass toward the drafting pond where we would fill up the buggy's tank. This burn show was winding down. The helicopter made a few reconnaissance laps around the fire, checking the lines. With the sunset, humidity rose and the fire lay down, making a few short runs in the pines but otherwise slowing to the speed of a walk.

The buggy lurched to a stop at the drafting pond, a man-made circle of upturned earth that had previously been used for watering cows. Fakahatchee Mike was unfamiliar with the shifting mechanism of the buggy and complained about it. "This here lever is sticky," he noted, pulling at the handle. I clambered to the back deck to help out with the drafting. Drafting water into the tank was a fairly simple procedure. Park the buggy right at the water's edge. Plop the black drafting hose with its heavy foot valve as deep as I could, using a shovel to prop it if the bottom was silty so that it did not clog. Flip a couple of switches on the pump and wait. In a matter of minutes, water came bubbling out of the fill pipe and we were done. The normal rule was not to run without at least a quarter tank. You might need that much to survive if you were surrounded by fire. Wet yourself down and pray, I thought.

"Darted Monkey." Fakahatchee Mike chuckled some more. He had a boyish gleam in his eye that indicated that my nickname would soon be common knowledge across South Florida.

I didn't mind much. It was all part of being on a crew. We tormented each other endlessly. Once Mike even caught a baby alligator and stuffed it onto one of the seats of our Ford crew cab for the next victim to find as he opened the door. I would get Mike back somehow and he knew it.

The slope leading down to the pond was angled steeply, and Fakahatchee Mike struggled to move the lever from reverse to park. "Click it out and then back in," our Mike shouted, indicating the gear shift, but it was too late. Gravity won.

Slowly, inevitably, the buggy rolled into the pond with us still on it. Water bubbled around us, covering the deck but not reaching the seats. I sat there for a minute, unable to believe what had just happened. We all started to laugh.

Taking the coward's way out, I abdicated responsibility, sloshing through knee-deep water and leaving the Mikes to explain. The last thing I heard was Jed's raised voice on the radio.

"The buggy is where?" he bellowed as I fled, gaining the pickup truck parked on the fire trail. I knew that this was a major mistake, one that would involve days of drying out spark plugs and cleaning brake shoes, but I couldn't help laughing. This day would go down in crew lore, the day Mike floated the buggy. Moments like these helped break up the concrete weight of sorrow.

I dreamed one night that Roger came back. He cupped a hand on my head, leaving it there for a moment. He never would have done this in real life; even though he felt like my brother, we never touched. In my dream I felt the heat and weight of his hand. I thought he was trying to tell me something, but I couldn't figure it out. "I have to go see Jen now," he said. I woke dazed, as though I had been working the night shift.

The next morning, I stood on a folding chair, taking down boxes of ping pong balls, still chewing it over in my head. Jen walked in with an armful of tools to rehab.

Her face gave nothing away. I hesitated before I spoke. Should I bring it up? Maybe it was better to stay silent, the way the guys did.

"Did you have any strange dreams last night?" I asked her.

She nodded. "I dreamed I talked to Roger last night."

I stared down at her, a thin blade of possibility working its way down my back, shivering in spite of the damp heat that permeated the work center. *Roger,* I thought. *Where are you? Are you with us still?*

Every morning when I woke up, I expected to hear the cough that had grated on me. I looked toward Roger's room and when Jim emerged, donning suspenders, I was momentarily surprised, until I remembered. *Oh, right. Roger's gone. He's never coming back.*

The women were tasked with history. We built a memorial garden, turning the dirt, copper taste of sorrow in our mouths. The guys didn't talk about it. They stared, inscrutable, at the scenery before them. They lived firmly in the present. If they mourned, they did it in solitude.

When we drove the swamp buggy out onto the refuge, I found myself searching for those places that Roger liked. The sweet orange tree, seeds from some logger's lunch that grew up into a shady spot where we stopped to gather fruit. A thin-needled Norway pine that Roger always pointed to if we were lost; the only pine of this type for miles around, it showed us the way home. Remember how he hung up the tree over there. There's the place we floated our ATVs when the water got too deep in that crossing, hanging on with one hand on the throttle, following Roger's lead until we gained solid ground. Here's where he took us one day on an exploratory hike, wandering deep through Unit 2 to seek out orchids. There was the faint path to a secret shack, left behind by the hunt club. Set on pilings, water flowed underneath it in a slow tide, and Roger had found it first and showed the rest of us. *Remember. Remember.*

Though it was a hot fire season, one of the hottest I could remember, it was tinged gray with sorrow. Roger strolled through my dreams more than once. I waited to see him hiking to the work center, shoelaces trailing behind him. I waited to see him wading out of the pineapple patch, cursing Bill for stealing all of the fruit for himself. I watched for him under the buggy, grease gun in my hand. He never showed.

As the years passed, other people died. Two helirapellers, high in the Salmon River breaks. Four people, a couple of them on their first fire ever, trapped in a box canyon in Washington. A woman, racing fire uphill in California. Each time it was a dart to the heart, something that wedged beneath the bones. Never forget, we said. But we did.

One October a firefighter I knew took his own life. This was nothing new. A dangerous month, October. Unless you could pull up stakes and follow fire to the Southern states, you wallowed in the troughs, the high abruptly gone. It got mean and cold in October, the snow line flirting with the mountains, any hope of fire long gone until June, sometimes July, sometimes never if it was a wet season. In

October, if demons already lurked in your soul, there was not much to stand between you and the gun.

There had been others that had gone this route, not by fire but by their own hand, but this was the first that hit close to home. It was a shiver in the night. There were so many ways that we watched out for each other on the line, but there was nothing to hold us up when we returned home. We scattered across the country and might never see each other again unless by chance we wound up on the same fire once more. At home, everything we tried to escape was still there, pinning us down. In October, there was nowhere to run. There were so many ways to die.

Ginger

Tonight my name was Ginger. Next time it would be something else.

Juls and I crowded the small trailer bathroom, getting ready to go to town. I didn't know for sure about Juls, but I was trying to become someone else for the night. I was proud of the biceps that allowed me to hold a growling Stihl chainsaw as it bit into the rock-hard flesh of a dead tree. I liked the strong legs and muscled calves that propelled me along a fireline, double-time, my pack bristling with thirty pounds of gear. Both Juls and I had worked hard to get where we were.

But tonight I wanted to belong. I wanted to be one of the women I saw in town, smooth skinned, feminine, delicate as shells, the kind of women men looked at as they passed on the street. Not me, my skin permanently smelling of transmission fluid and saw gas, my fingernails stained a perpetual black.

I watched Juls run a comb through her long hair, recently colored a shade of cinnamon. She had recently joined the crew, bringing our female total to three. Juls was confident in her own skin as if she always had been. She didn't need fire to feel that way. As we cut slash pine in

Burn Unit 2, removing the dead trees, called snags, prior to our next prescribed burn there, she strolled easily, her saw in her gloved hands. These dead trees, victims of a fire that got too hot, terrified me. They resisted our saws, forcing us to pound wedges deep into their unforgiving bark to make them fall. They twisted in their falling, starting to sit back on the saw bar. There were so many ways it could go wrong.

I was a timid sawyer, darting back and forth on either side of my face cut. My back cut wavered dangerously, and I leapt out of the way as the tree thundered down to earth. I was a reluctant sawyer too, frozen in fear by what could happen: barber chair, saw kickback, mistake. Injury, deep welling of blood through chainsaw chaps, a leg mangled, death.

You had to commit to the cut, though. One time Steve had gamely kept with his back cut, waiting grimly for the tree to fall before shouting, dropping his pants, and slapping ferociously at the swarm of fire ants that had been biting him during the whole process. Tree falling was not for the meek.

Roger used to help me with my hung-up trees, tiptoeing over to laugh at the branches stuck firmly in their neighbor, the result of not sighting the fall line correctly, or the chainsaw bar I had pinched, the log closing firmly on the chain. "What were you *thinking?*" he would howl. Always he would talk me through the tricky extrication. He was gone now, and I was on my own.

Juls took to tree-falling dead snags like she took to everything else. She could tie knots, weld, and drive the unwieldy swamp buggy with ease. She often watched the rest of us with impatience as we struggled with a task that took her minutes to complete. Her hair in a long rope down her back, she was creeping up on what could be considered the wrong side of forty; she took up space in the world with her laughter. She was not afraid to be heard.

Outside the bathroom, Wayne, an AmeriCorps volunteer conducting a plant regeneration study, sat glumly on the couch, waiting for us to finish so he could get to the washing machine. He was deep into endurance laundry, endurance because he let it pile up so long that it took hours. The washer rattled away near the rusty bathtub, fogging the mirror with its hot breath. The water, pumped directly from the swamp and into our leftover FEMA trailer, tinged anything white a dingy yellow. It turned my hair to straw. Rumored to contain high levels of mercury, it felt like slippery toxicity sliding down my skin.

At least he was close to the only coolness we had in our shared trailer, an evaporative cooler, nicknamed a swamp cooler, that chugged ineffectively over the sink, creating an arctic blast that covered one square meter. It was winter in Florida but you wouldn't know it. No semblance of cold air made its way down the single-wide halls with their itchy, indoor-outdoor red carpet and into our cramped rooms. At night I sweated, one sheet draped over my body, the window cranked out as far as it would go, hoping for a breeze. The only relief was standing beneath the cold shower, imagining the silver glow of mercury lighting me up.

Wayne was big and tawny, a swamp boy from upstate. I didn't know much about him. He wouldn't be here long; he was just one of the students that came and went. He was working on some complicated thesis involving plants that came back, or didn't, after summer and winter burns. We had helped him with his study, wandering through the prairies collecting small succulent samples of unknown plants. He tried to engage us in conversation about this, because certain flowers, he said with enthusiasm, only came back after a cooler summer burn, when water surrounded the prairies. Didn't we care about that?

His face fell as we shrugged. Honestly, we didn't, we told him. We just liked fire, no matter the season. We were not versed in fire ecology; we were simply the bodies it took to put fires out or to light them, depending on what someone else determined needed to happen. There were times, though, when the tall sawgrass whispered against my legs as I walked, that I wondered about it all. I watched where the fire had been, skipping inexplicably over some chunks of land, lingering on others. I saw what fire left, a scatter of white ash where a tree had burned hot, the fields full of the twisted bodies of snakes too slow to escape. I was overcome with the mystery of fire then. It was both a living thing and a contradiction, a push and pull at the same time.

"Jim," I hollered down the hall. But Jim had retreated into his room for the night and wanted no part of town. Wayne didn't want to go to town either. He was content here in the swamp, miles from any other lights. He viewed us with a mild curiosity. Heaping baskets of camo pants and formerly white T-shirts surrounded him like armor.

"Almost ready?" Juls asked, adjusting the straps of her tiny sundress. I leaned close to the mirror, swiping a rag across its opaque gaze. A ring of poison ivy braceleted my forearm. My hair, the ends frizzy from catching on fire yet again weeks earlier, wrestled the brush.

I felt all wrong. There it was, every flaw laid bare under fluorescent lights. My shoulders, like a linebacker's, too large from decades of hauling packs and wrestling blackened wet inch-and-a-half hose into a manageable carrying bundle. My toenails, destroyed from repeated hikes down steep mountains. I was no mystery, no local woman with blow-dried straight hair and dewy skin. I had to face it: the only time I felt right at all anymore was fighting fire.

On spotting Juls and me cutting trees, drivers on the highway had called the police, suspecting poachers on the wildlife refuge. Our hair was tucked into sky-blue hardhats, our faces shrouded by safety glasses. We carried big saws, shoved wedges and falling axes into our back pockets, and buckled our legs into drab green chaps. We felled trees that could spit firebrands across our lines, stopping only to check the holding wood to see how cleanly we had cut.

Wood chips tumbled into our bras. On breaks we coated our skin with a slimy, perhaps toxic yellow substance marketed as preventing poison ivy infection. Sometimes it worked, sometimes it didn't. Sometimes we rode, holding on with white knuckles, in the back of a rickety trailer, spraying harsh blue chemical on invasive Brazilian pepper. Jed drove the swamp buggy in shirt sleeves, unfazed. "Already had my kids," he bellowed as the poison settled in a cloud over all of us.

When I saw the city firefighters in town, they didn't even recognize me. I walked past, my hair unshackled, my legs strangely bare. They only knew me in my flame-resistant shirts and baggy pants, the shell that I became when I fought fire: androgynous, neither woman nor man but something in between. Sometimes it felt as though I occupied two worlds, firefighter and not, but never completely lived in either one. Before Roger died, I had never felt such a gap. Fire was what I chose. Now, doubt had crept in.

So I would be Ginger tonight, to Juls's Mary Ann, even though it should have been the other way around. Juls was the Ginger type, not me. I was too shy in my city clothes, too visible, but the name switch was meant to be funny. The men we met in the clubs in town would not even understand that everything about us was made up. Our outrageous names, the courage I painted on my skin, bold red lipstick, peacock eye shadow, a mask that I wore. Tonight I would visit the woman I could have been, if I hadn't become a firefighter instead.

The perpetually blond trust-fund men would dance awkwardly with us, conversation impossible with the pounding techno music. "I'm a biologist," I screamed over the noise, and they seemed to either believe it or not care. They wouldn't see the flat calluses on my hands from gripping fire tools or the cold steel of a pull-up bar. They wouldn't feel the heat that I thought I exuded after years on the fireline, the decades of smoke I breathed out with each breath. I could be anyone. I certainly wouldn't be the same woman I had been earlier that day, cutting trees.

When the men weren't looking, we slipped away into the darkness. Though it was Sunday for everyone else, our shift started early tomorrow, and it could consist of many things. A fire could mushroom up out of the Blocks, requiring us to chase it, driving our swamp buggy on the dozer tracks laid fresh and bleeding into the heart of the forest. We could be sent out to California or Idaho to a big project fire with just enough time to get to the airport. It could be a work day, us swinging our axes to clear the fire trails. We needed to be fresh, and so like Cinderella we vanished, shoes in hand. By midnight we would be back in the trailer, hoping for a breeze, our hair stale from a different kind of smoke.

As we pulled out of the club parking lot, I realized that I had lived on the edge of this town for several seasons but knew nothing about it. I didn't read the newspaper. I didn't know who was mayor. I couldn't join any clubs or take any yoga classes. I didn't have a library card. My license plates were from a different state, a state I had not lived in for years. I didn't know anyone here who was not in the seasonal world. In a life where I could get a phone call summoning me to a three-week fire assignment and be out the door in an hour, I couldn't do those things.

"It's too late for me," Juls said without a hint of self-pity, when we talked about babies. Instead she talked about all the things she could do: join a trail crew which didn't pay much but the air was clear, go to Europe, build a house. I had no doubt she could accomplish any of those things.

Juls and I drove down Alligator Alley, the ribbon of interstate belting Naples and Miami across Florida's wide bottom half. We were stone sober, because we needed to have our A-game on the next day. One misstep and it could mean someone, maybe us, dying. This was a reality we always shouldered.

If we kept on driving we would end up at the other side of the state, as far as we could possibly go. We would be backed up to the edge of the country then. I sometimes wanted that, to realize that I had driven as far as I could.

There would be no room to keep searching. Maybe that would make me stop, finally. I didn't say anything because Juls might not understand. She could take fire or leave it; when it stopped being fun she planned to quit and do something else. Juls was the type who always landed on her feet. Sitting in the passenger seat, I realized that I could learn this. I could not be Ginger, but I could figure out how Juls moved through the world, unafraid in her stride.

We turned off onto the dark ribbon of Highway 29 and negotiated the tricky left-hand turn, easy to miss in the darkness. Next door the exotic animal farm was silent except for long growls from a lonely tiger.

Back at the trailer Wayne had gone, leaving a faint scent of Tide and the lights blazing. His camp trailer was dark and so was Jim's room. The mirror was still fogged with dryer heat and I hastily splashed water on my face as I turned back into the person I had tried so hard to become. It was true that it felt much more comfortable than being gussied up at the club. There, I felt like I was pretending. I didn't know how to be in that world. Sexy, funny, one of the guys? The choices were bewildering.

So far fighting fire was the closest I had come to the person I wanted to be. I didn't want to lose this person now that I had finally found her, but what I did know was this: the fireline was a tenuous thread. You were dependent on the shaky combination of lightning and wind. You were foiled by the mysterious workings of your knees, the constant scrape of bone against ligament. You were only as good as your last pulaski swing, and you were judged by your speed, your strength, and your desire. It was easy for any of these things to fail. You were always racing age and mortality, and you made hasty bargains to stay in.

You gave up relationships and summers, plans and babies. If you were a woman, you had to prove yourself on every shift. You ducked behind trees to pee until you just didn't care anymore if someone saw your pants around your ankles. You worked harder than you ever imagined you could, until you reached a kind of plateau where nothing hurt anymore. You walked the dangerous line between joy and despair a thousand times a day. It was both the best job you ever had and the worst. It could seesaw wildly between the two on any given shift.

Most days it had been the best, for reasons that were hard to articulate. Most days I worked hard to think that I was still twenty-two and could climb mountains faster than anyone I knew. I tried to forget that the fireline had an expiration date, that it would take everything it could from my body and leave me with nothing but memories. *It's still worth it*, I whispered to myself, a kind of mantra to hold close when things got tough.

This lasted until I lay in my single bed, down the hall from the room that used to be Roger's. Even though Jim occupied it now, Roger's presence was indelibly imprinted in the walls. As much as we cleaned, it would never come out. We tried to bring him back with our butterfly garden and our raised glasses, but nothing worked. He was never coming back. There was only sorrow that lay heavy on our shoulders.

I went everywhere the year after Roger died. I went to Texas, to California, and to the Northwest. Fires burned all year in a long loop, starting in Kentucky in winter and finishing up in the Cascades in the fall. The fires were the same but I wasn't.

A rookie leaned against his red bag on a Texas airfield. "I'm in this for the glory," he boasted. I remembered the days when I had believed that too.

"I don't think we should have relied on that safety zone a mile away," I said to the crew boss on another fire. He swung around to take in the rest of the crew. "Anyone else not feel safe?" he asked in a tone that discouraged responses. There was only disapproving silence. I was speaking against the code.

I hiked through the burning woods, seeing danger in every flicker of flame. Didn't they know? I wanted to scream. Didn't they have any idea of how fast it could all go wrong?

It was easier to think small, to narrow my world down just to the meander of a fireline. I tried not to think past the next hour as I leaned over my pulaski. By now I could tell what would be easy and what would take the crew hours, shock waves running up my arms with each swing. In the deep forest of the Northwest the duff piled up, cake-like layers of pine needles and cones and twigs, making it difficult to get down to mineral soil. Dense forest meant thick-limbed trees, each branch of which we had to chop and throw in large armfuls to the green side. It could occupy every corner of your brain, this fighting fire, leaving no room for anything else.

There were times when I forgot, as if I had slipped through an invisible curtain back to the time before. There was the jump of my heart at the sight of flames pouring like a river through the forest, the sturdy weight of the pulaski in my gloved hand, the pungent smell of fresh smoke. On a night shift in Kentucky Jim leaned back against his fire pack. "It doesn't get better than this," he quipped. We were all cold. Trees fell with muffled thumps somewhere in the darkness. A house squatted on a hill nearby and the occupants had offered us coffee, but the crew boss would not let us leave the line to get it. He offered this arbitrary rule with a shrug.

We had been digging line all day in a hardwood forest, leaves falling onto our shoulders like rain. Anti-government residents set fires below us, seeking to scare or trap us, we weren't sure which. We were warned that there were feuds in the area and that bullets might whiz past our heads.

The trees etched a darkened sculpture in the sky. I huddled miserably in my inadequate space blanket, waiting for the sun. I looked from face to face, all reflected in the amber glow of our campfire, and I saw what I loved about fighting fire in each of them. Toughness. Courage. Love.

Others traded in their fire boots. Steve, my old Florida foreman, called one night with the enthusiasm he had previously reserved for a running fire in the slash pines. He had fallen hard for a curly-haired woman and was moving to a state without a lot of fire. Jed contemplated retirement. Now when I went on fire assignments the faces were those of strangers. I was the fire grandma, the one you went to when you needed advice for the lovelorn, moleskin for your blisters. "Are you married, do you have kids?" everyone asked, and I shook my head: No. No.

I wanted to say: I have this. I have decades of this, memories grown a little hazy with time, the hand of a friend giving me a lift up from the stony ground during a twenty-hour shift, the eyes of a man long gone meeting mine as we stepped into a prairie as easily as swimming. I have the story of a woman who learned to love fire and grew to like herself along the way. Instead I reached for my pulaski. "Close the gap!" I yelled. I headed up the mountain as fast as I could go.

On Storm King Mountain

I knew it was time to go to the mountain. There was only one mountain I meant when I thought this.

Storm King. Its long shadow had loomed over me for over a year, only occasional slivers of light penetrating the darkness. Others had gone and returned, their faces somber. It was hard, they said, a bolt to the heart. But good, they said in the same breath. You should go.

I was afraid of what I would see. I had looked through all of the pictures, the crumpled silver of fire shelters on the burnt hillside, the lone tree, its arms bare against a mean sky. The smoke column, boiling up like an atomic bomb. I thought it would be a terrible place, full of pain.

I was afraid of what the mountain would tell me. I was afraid of what it knew. I wondered if the screams of those who died could still be heard. I wondered if it would be possible to breathe.

But Storm King Mountain was healing, faster than any of us could. I parked in the subdivision they had saved. Nobody was around and I wondered about the burden those homeowners carried, different than mine but perhaps just as heavy.

I climbed up the narrow trail to the place where Roger died, one year later almost to the day. Green shoots poked impudently from skeleton bushes. Birds chirped from blackened trees. Deer had left meandering tracks along the lean trail that the firefighters used as their western line. Far below, cars droned by on the interstate, the one that some of the survivors managed to reach during the firestorm. The others did not, caught in a race with fire that even they could not win.

White crosses leaned into the hillside to mark where each person fell. I could see them standing out against the black as I climbed closer, each one staggered in a close group. Visitors searching for meaning had carried souvenirs to place at each one—a set of skis, cans of beer, a ball cap.

I knelt by Roger's cross. He was not very far from the top. Ten minutes, fifteen? He was close enough to see it from where he took out his fire shelter. At some point, he made the decision that he was not close enough, that even his skinny legs could not move fast enough to get to safety. Remembering the fire I had fled as a rookie, I could imagine the chaotic scene, the push of wind at their backs, the darkening sky, embers swirling just ahead of the main fire front.

Before those last seconds, fire had been boiling up from below, unseen. A combination of kiln-dry oak and winds pushed by a cold front blew a casual-seeming fire into an inferno. Roger and the others were building line below the ridge when they realized something was wrong. People already on the ridge lingered until almost too late, yelling for those below to drop their tools and run. The onlookers waited in vain, finally dropping over the other side of the mountain as the fire crested the ridge and hurtled across like an unbroken wave. They ran for their lives, stumbling and getting back up again, their fallen comrades lost forever on the other side of the mountain. The runners barely made it alive, coming out onto pavement, safe, hoping against all possible hope that their friends had survived. In the pictures the survivors had taken before they fled, I could see what the doomed firefighters could not: a wall of flame coming up the mountain toward them.

I tied a piece of bright pink flagging around the base of the cross. This was how I used to keep track of Roger in the woods on all our red-cockaded woodpecker surveys. As we walked parallel to each other, we would look for the ribbon we both wore twisted through the bands of our ball caps. Sometimes it was hard to stay on a correct bearing, because we had to detour around swamps and fling our bodies over vine mats. Going too far off course could mean that we would miss a colony.

Red-cockadeds had never been spotted north of Alligator Alley. Since they were only found in about three percent of their former range, it was doubtful they ever would be seen on our refuge, and most of the crew thought this was an exercise in futility. But we were commanded to survey before each burn, and so we did. Most often we emerged near the waiting buggy, clothes torn, hair filled with sticks, a raging case of poison ivy on the way. Sometimes we never even glimpsed one solitary pine, the trees the birds needed. There were snakes and wild pigs and fire ants. But I still believed. There was always that chance of stumbling into an open stand of thick-bodied pines that just might hold an endangered bird.

Roger believed too. Sometimes the rest of us would be out on the buggy already, impatient to get back to the work station for showers and beer. "Roger," we would scream into the woods. "Get your ass out here!" He would still be out there, head tilted to the sky, hoping to see a small woodpecker.

I had been tough for so long that I thought I didn't know how to be any different. But this incomprehensible thing hit me like a punch in the stomach. I lay on my back at my friend's cross, harmless cumulus clouds passing overhead. I watched the slow cartwheel of earth and heaven, feeling the mountain's steady heartbeat beneath me.

I thought about how I was living my life, the way I flowed across a place without really touching it. For the first time I wanted my life to have some meaning besides the pursuit of adventure. I wanted to come home to someone and to a place where I was known. I thought about Roger and how I just expected to see him again, the way I expected the sun to come out after a storm, the way I expected to see all my fire friends again. Suddenly I wanted to be anchored to a place, somewhere that I could see all the seasons, not just one. I wanted to live somewhere other than a bunkhouse. I wanted to imprint myself on life, to make a mark saying that I was here.

I walked up to the top of the mountain, an easy walk without fire below, to the saddle where some of the firefighters had seen the crew below them and hollered for them to run. One of them had said he and Roger had locked eyes, and Roger wore a concerned expression. He was the last person living to see Roger alive. That was the last of the crew that they saw before wheeling and making their desperate run to survival. It was a run I had taken before, in a different place.

Far away in the long distance, ancient coal seams smoldered. The site of the Vulcan mine had been burning underground for over a century. The smoke would outlast all of us. Nobody was up here but me. It seemed like a long way from home, even though I had no idea of where home was or could be.

I stayed on the flank of Storm King, this beautiful burnt mountain, for a long time. I hoped that some kind of meaning would seep into my bones as I lay where my friend had died. I pressed my face into the ground, smelling the trace of burnt ground and the promise of new life. It smelled good and alive, all sorts of invisible organisms busily working to repair the soil, a thriving construction zone. Somewhere beneath, seeds that had waited for hundreds of years were working their way to freedom.

Alaska

Alaska was the place you went when you had tried everything else. I went there because I wanted to slide inside the ocean's skin, to feel its endless push and pull like a breath held and then released, a long deep sigh throughout my bones. I wanted the calmness I imagined lay below the surface of the sea. I wanted to melt into a country, to be absorbed by it. There were many reasons I went to Alaska—a romance gone wrong, a job grown too sedentary—but mostly, in Alaska, I believed that I might be able to stop thinking about fire. About Roger, wrapped in the dissolving remains of a fire shelter as the fire closed in, taking his last breath. I thought that I could grow into the woman I was trying to become, a woman tough yet open, a woman who did not need fire to make her whole.

I wanted to wean myself from fire. The only place in the country that had not been shaped by it, a coastal rainforest archipelago tempered by the sea, seemed to be the place to go. I took a wilderness management job, one that kept me at a desk and on the ocean in a kayak. I was learning to live without fire, finding the balance between tide and beach, the paddle in my hand not unlike the pulaski. I liked the steady sameness of the ebb and flow of the water, estuaries emptying themselves out twice a day, hours later the return. I taught others how to stroke through head-high waves and where the uncharted rocks lay hidden beneath our skinny fiberglass boats. I liked this life of rain and sea, but fires still happened and the crew bosses stormed my cubicle, pleading for my help. "We need you," they said. "You're the only one with the right experience. Just a few days?"

Here in Southeast Alaska it could rain six feet a year. Here, the occasional abandoned campfire burrowed deep, hidden and flickering down through the mounds of duff to the beach cobble. It was dry under the big trees and the fires smoldered for weeks, lazy and unhurried. Because these coastal forests were not fire adapted, fire took the land back to primary succession, turning the soil bare and hard and new, something it would never have been. We pounced on them, saturating them with salt water to put them out. In this harsh climate, it took a thousand years to grow an inch of soil. This country, changing rapidly with the effects of a warming climate, did not have that much time.

But the Alaska Interior, hundreds of miles to the north, was a different story. Even though it was located in the same state as the soggy coast, this was a place of fire. The interior burned white hot. There were so many fires that most were not fought until they reached spitting distance of villages. They sprawled out over hundreds, thousands of uninhabited acres. It was fire in its purest form, fire allowed to do what it had always done. I flew over fires that stretched over hundreds of acres, bright flame chewing through tundra, nobody fighting them. Winter snow would put them out, or autumn rain.

In the Interior, I walked through a burning forest, watching the black spruce torch out above my head. This fire was a costly mistake—some local girls, blueberry picking, losing control of their cigarettes or their warming fire, nobody was sure which. Although the nearby village knew which girls they were, there was no attempt at censure. They saw fire differently here. To them, fire happened. It was a living thing, part of the landscape. It always had been, up here. Fire was not an enemy or a friend. It just was, like the endless winter snows or the spring river break-up. I wondered if I had been seeing it all wrong.

Despite the fact that fire was now like a stranger I had once loved, I still fell into its rhythms as I was pulled from my desk job in an unusually active season. I loved the musical names of the villages we fought fire in, names like one long poem: Aniak, Kalskag, Talibiksok. I loved our camps, hardscrabble black plastic and drooping tents. I loved how we were left out on the tundra—one crew, a shotgun, and a hundred thousand acres—to fend for ourselves until the fire was out. On one fire I ended up with a handful of old-timers like myself. We took stock of each other.

All of us were veterans now, twenty years or more under our belts. These were men who couldn't quite let go either. Alaska Doug, his curly hair frosted with silver, was facing hip surgery, a fireline injury exacerbated by decades of walking the line. He walked with a little hitch in his step and a grimace on his face when he thought we weren't looking. Another guy was talking about retirement, even though that felt like something for old people. We gathered by the warming fire, wrestling with our decisions.

"This is my last season," we all said, but we kept a part of ourselves silent as we said it. It felt like none of us could imagine a season without fire.

There had been other wildfires in the Idaho mountains since Roger died. There had been other fires in the Gambel oak of Colorado and other fires in the Southwest desert. The cycle continued, as it would continue long after all of us were gone. Other crews would hike the line with their tools for generations to come.

Juls, long-haired, strong, beautiful Juls, had quit fire. For now she worked the seasonal life on an island park, running a boat and swinging a hammer. Florida Mike and Jen, now married to each other, ended their fire days long ago. Jim worked on a ship, as far from fire and snakes as he could get, and Jed had retired with little fanfare. They all seemed content, the letting go less hard than they could have imagined. One of the trail crew Dans had suddenly, inexplicably, died of melanoma. Others had melted away somehow, their whereabouts unknown: Little Mike, Chester, my old fire buddy, Doug. Whether they were still on a fireline was anyone's guess.

Others carried the torch still.

I thought about Roger, gone now for ten years. A lot could happen in that time. In ten years, I had moved, four times. Babies I knew then had grown into long-haired girls. I had grown too, moving incomprehensibly from young to not so, knees turning cranky, wrinkles fanning out from my eyes. Ten years seemed like a long time.

In the places where Roger and I used to walk, the water flowed in an endless sheet to the Gulf of Mexico.

Cabbage palms crowded the sky. The panthers slipped in and out of the blanket of night. None of this had changed. To most of the world, ten, fifteen, twenty years was nothing, a blink.

Still, I carried Roger's last fire with me wherever I went, along with the dusty ball cap he used to wear. It was a sleeper, no big deal at first; no different than the others we used to fight together. Surely he was not thinking about dying then. After all, he had just fallen from the sky on the strings of a parachute. He used to say he liked it because he felt closer to God in that moment.

Because I was on another fire, a few miles away, I knew that where he was the sky was cloudless and strangely beautiful. There was the satisfaction of watching my pulaski cut through the layers of earth. There was that profound indescribable feeling that all firefighters know when walking the line. There was nothing to indicate that things would go so wrong.

The mountain Roger died on had burned once since. It would burn again. I could not stop the Gambel oaks from growing, the lightning from coming, the years from passing. But I knew my days on the fireline were numbered. I knew it this time, deep in my bones in a way I had never known it before.

Things had changed after South Canyon. There were no more forty-eight hour shifts and three-week assignments were cut to two. Spike camps were no longer the hardcore places that Dave and I endured so long ago in Wyoming. Human resource specialists appeared in the camps, pleading for everyone to just get along. Crew bosses were encouraged to be less old school; they were supposed to give us full briefings. Anyone on the crew could voice misgivings. Much talk was given to refusal of risk. When Roger had been alive, crews were expected to take any assignments without question. As crew members we had been expected to follow the crew boss with blind faith. This, some said, had led to the deaths. It was implied that the fallen had brought it on themselves.

There was a scattering of women in top positions now, higher than I had ever been able to achieve, women who supervised entire divisions—not many, but some of our second generation going on above the crew level. They carried their gray hair and their experience proudly. It was good to see them out there. I was proud of them for sticking it out, for making the hard choice.

The fire orders were revised to reflect the change. Before, the first one dictated that we must fight fire aggressively but provide for safety first. The new version had us dropping back to size up the weather and the fire first. Actually fighting the fire came last. This was the way we had always done it, but now it was put into words that made sense.

Some of us sat around our desks on our off days talking about the old days. Michele and Kent had started around the same time I had, although we had never met on a fireline. The young firefighters eddied around us as if we were something dangerous. To them, we were loose cannons, representatives of the days when anyone could fight fire, long before the massive fire hire of 2000. This hiring, a reactive response to a multi-state firestorm, meant that it was mostly those in full-time fire positions who went out on fires. The call-when-needed crews were mostly put to work mopping up, if they were used at all. Nobody wanted to return to the days of forty-eight hours without a break, twenty-one day assignments. Nobody but us.

"They have bottled water, not cubies anymore," Michele said.

"They fly crews out of spike camps to take showers," I added.

"Well, I'm just glad we got to be part of it," Kent said, putting a good spin on our history.

With so many years behind us we remembered it all as being flawed but perfect, a crack in a glass ball. Memory was tricky that way. Everyone's firelines were the slimmest and hardest-won, their fires the most spectacular.

But nobody could deny that the fires now were different too. With the gradual warming of the West, fires were beginning a full month earlier than when I first stepped on a fireline. They were starting higher in elevation too, in colder, wetter places where the trees had no defense. Researchers posted studies showing that right around the time of the Yellowstone conflagrations, fires became more frequent and burned more acres. Since 1987, my first year on the line, the average length of the fire season had increased by seventy-eight days.

These fires were bigger and more intense. Crews were pulled off the lines every day as the fires blew up, over and over. Assignments were changing to mop-up or protection of cabins, hotline reserved for smokejumpers and hotshots only. Or there were just big air shows, the ground deemed too dangerous for anyone. Costs soared into the millions, with national overhead teams bloated to one hundred people in air-conditioned yurts. Gone were the army tents, the flapping yellow tarps. Fire was now big business.

On one fire in Washington state I ran into a vaguely familiar face. We started the old guessing game.

"Were you in Idaho in 2000?"

"Nope. Oregon. What about Canada? Did you get to go there that year it burned?"

"No, I was in Montana that year. Rock Creek maybe?"

"Not me. I know! At the dip site on the Lochsa?"

"Nope. Wasn't me."

We almost gave up when it finally came to us. There, decades later, was my first crew boss, Alec, the one who despaired of our wobbly-kneed, inexperienced crew.

"I didn't even think you liked fighting fire," he said.

I shrugged. "Yeah, well. I guess it stuck." I could see in his eyes how far I had come.

We grinned at each other. Helicopters lifted off the field behind us. There was no time for catching up. Someone called his name; he had to head up to H-1, one of the helispots. I had to leave too. "See you on the big one," I said, knowing we might run into each other again, or might not.

A crewmember, Rich, came along with a swivel and some nets. We were going to the line to set up some fifty-gallon blivets. "How do you know that guy?" he asked.

I swung around; Alec had disappeared, absorbed in the crowd of firefighters waiting for helicopter rides. There was so much I could have said about that fire. Our reckless run to safety during the blow-up. The hotshot woman with eighty pounds of water, the one I had modeled myself on for decades. I looked at Rich's incurious face. There was a job to do, a helicopter waiting. There was no time to dwell on the past.

"It was another fire, a long time ago."

I grabbed my pack. "We'd better do something, even if it's wrong."

Rich shrugged. He was in it for the money; nostalgia meant nothing to him. He settled his flight helmet onto his head and rocked back on his heels. "Time to fly, Whiny Baby," he said.

"I am not a whiny baby!"

Rich lifted an eyebrow. The night before we had thrown out our sleeping bags in a field of head-high brush, finding old concrete helipads to escape the spiny weeds. The helicopters we would be working with had not yet arrived, and the silence was only broken by the snoring coming from Rich's direction. In the morning I had stomped over to his location to demand he move farther away the following evening. In that time, though, a half-dozen helicopters had landed in the field, which had been arduously cut by a guy on a riding mower. Beetle-like helicopters squatted everywhere, each attended by their maintenance crews. There was no longer space to spread out and sleep.

We marched out to our helicopter, which was staying hot—rotors turning—for a speedy departure. Rich and I settled into seats facing each other. His blue eyes crinkled with delight at having bestowed a fire nickname. We lifted off from the helibase, fire camp quickly receding from view, a line of mountains replacing it. The pilot spoke into his microphone, pointing out the helispot we were aiming for, a clearing carved out by chainsaws surrounded by dense forest. *Hundred-foot longline*, I thought, calculating the distance of line the crews back at camp would need to hook up to the helicopter in order to clear the forest canopy if it was slinging in supplies in nets to us. As we descended straight down, a fine cloud of dust mushroomed from the ground. *Water drops*, I thought. We would have to call for one of the bigger helicopters to drop a load to keep the dust down. By now this thinking had become automatic, a series of calculations that ran seamlessly through my head.

Rich and I scrambled out of the helicopter and held our thumbs up as it peeled away. We were only supposed to be here for the day but we had come prepared with overnight gear. You never took helicopters for granted. Things broke, or the fire took off somewhere else and you became a low priority. You were always at the whim of engines, wind, and fire.

I ran into someone else, too, someone I had not thought about for years. Fifteen years after I last saw the smokejumper on an Idaho mountain, it was another a catastrophic summer in the same place. The fire season in Alaska's Interior was over, the rains finally coming across the mountains in thick bands, but down here in Idaho it was another story. I was sent down from my wilderness ranger job to help out—even faraway Alaska was drained of firefighting resources. The combination of beetle-killed trees and dry lightning was a deadly mix. The forests literally exploded, their columns towering thirty thousand feet in the air. Evacuation orders locked in place, residences threatened. One day I walked into the briefing tent and there he was, my smokejumper from so long ago.

The first thing I noticed was how small he was, barely taller than me. Over the years I had built him up, used the excuse of this relationship to stay out of others, to hide behind walls. In the telling and retelling to myself he had grown in stature until he towered over me. I could not believe that this small man had cast such a large shadow over my life.

Instead he was short and slight, still built like a jumper, lean and strong. I probably could fit into his clothes. He sported new wrinkles around his eyes. They suited him.

He was just an ordinary man after all.

He smiled, gave me a hug. It was awkward, two former combatants standing in frosty grass, the fire camp swirling around us. It was not a place for heartfelt apologies or recriminations. I didn't want either. I wanted to close this book. I wanted to walk away.

We made small talk about fires we had been on, people we both knew. *How's the General? Seen Buck lately?* We worked our way up to Roger, slowly, testing old wounds. I said that I still missed him, that people who remembered him were getting harder to find. Instead there were blank stares from the rookies, too young to know, too in love with fire to care.

He agreed, his face somber. "It's good to see you," he said. I realize that I felt the same way. Mellowed from years of solo travel, I understood how it was for him, the push and pull of wanting a footloose life to follow fire, the desire to have someone by his side. Two sides of a coin, wanting both but settling for one.

People were waiting for him, and it was time to go. His division was still uncontained and a big wind event was supposed to come through.

"Keep one foot in the black," I said, the mantra of all firefighters. The black, the place fire had passed through and left behind, was the safest place to be.

He hesitated. "Stay safe." We parted, walking in different directions. I didn't look to see where he was going. As I headed back to the helibase, I realized something I had probably known all along: I had clung so tightly to the smokejumper not because he was anyone I couldn't live without, but because I wanted what he seemed to have. Not just fire confidence, though he had that, but confidence in general, something I had sought my entire life. Now I had that confidence, and I didn't need to chase after it in someone else. I didn't need to be someone's reflection anymore.

A weight I hadn't known I still carried lifted away.

A year later, back in Alaska's Interior, I walked with Billy, an old hand at this type of firefighting. I was the IC, and he was a crew boss, but we were equals in time and experience. His hair twisted down his back in a long braid. His eyes were deep and, I thought, wise. From a village deep in the Interior, he was in his element. This was all he had ever done, since the age of sixteen. His father had done it and his brothers and his kids did it too. Fighting fire was his livelihood, a few dollars squeezed out of an ephemeral fire season that usually lasted from May to July, with a hope of being sent south if it was bad enough in the West. He hoarded his money in order to survive the long winter.

While we sized up the fire, his crew rigged up a cozy hooch with parachute cord and black plastic. I struggled with mine, never quite getting it right, remembering the old-timers back in 1987, telling me how to make this kind of shelter. The crew laughed, watching, but didn't offer to help. Out here, as always, you carried your own weight. I finally got the shelter to stay up, but it sagged and provided little barrier against rain.

They peeled back layers of permafrost to make a homemade refrigerator. They built a rack for their tools. They carried a pot in one of their packs and brewed strong-smelling coffee when we took breaks. They scattered all over the fire, not needing any orders. They were naturals, in their element.

Billy and I checked the Mark Three pump that droned at the water's edge. We looked in on the crews, crawling into the interior of the fire with dirty hoses. Everyone laughed and joked despite the sloppy work.

An onlooker appeared, holding a dead porcupine by the tail. Dinner, he told us. He was only one of the locals who trickled up to watch us. We were the only show in town.

We walked on toward the place where someone had seen musk oxen a few hours before. They were gone now, the tundra stretching empty in the waning sun. I talked about Roger as Billy and I slowly walked, our eyes automatically scanning the fire for trouble. I talked about how I wrestled with what had happened to him and how I both loved fire and hated it now because of it. Billy just nodded, listening. His life was on the edge in his own village; people died at the hands of the wilderness all the time. He accepted this with equanimity. You go on, he seemed to say. You accept what is.

The second day of this fire, a young man had fallen off the barge into the nearby Kuskokwim River as the boat steamed toward the village. He was never seen again. This only punctuated what Billy implied. I started to think about it that way; life as an unending cycle, sort of like the fire that swept across Alaska's Interior every few years.

We sat in the soft tundra for a while, watching the fire. It had dropped down from the trees and was giving up the fight. The crews had lined it well. We had caught it in time and rain was in the forecast. In another day, maybe two, the rain would pound us with ferocity. It would flatten our tents and whip our black plastic into a crazy dance. It would seep into our boots. It would put out the fire for good.

The helicopter pilot hailed us on the radio; he was dropping in with the supplies I had requested. As I pulled myself to my feet, a familiar twinge in my left knee reminded me that I probably should not have put off surgery again. "You're so young to have such bad knees," the doctor in town had told me, shaking his head and recommending surgery. He didn't say it, but I knew: I had to choose. It was a constant refrain, growing louder, the unregulated beat of a heart.

"Better get moving," Alaska Doug said on the last day of the fire. He winced as he moved from our resting position. We sat up from our blueberry patch and hoisted our tools. He pointed south. He would stay on the fire's cold edge and I would step into the interior, checking for smokes. Dave would parallel me. It was just like the old days when Roger, Jen, and I walked in separate lines. If I closed my eyes a little and let them blur, the two men I followed disappeared. I was back there with Roger and Jen in the tall sawgrass while a prairie caught fire behind us. We were walking into the rest of our lives.

When I thought about fire I would always think of times like this: the three of us walking side by side, heading into the dense smoke. The three of us fading into the three of us that used to be, all the fires that had ever been dissolving into one fire.

The farther away from our blueberry hill that we walked, the harder it would be for anyone to see us. Even our yellow shirts would fade from sight. Soon we would all become one thing: the fire, us, the tundra. We were so entwined that there was no separation. It would be impossible to tell where we began and where the fire ended.

Epilogue

I am fifty years old and I am climbing Storm King Mountain on the twentieth anniversary of the fire. I have lived twenty more years than Roger ever will. I have moved countless times. I have been married and divorced, and now I am married again, and I hope that this time it will stick.

I am fifty years old and Roger's father, Wally, is thirty years older than that, but he hikes gamely behind me, determined to reach his son's cross one more time. We hike in a line with the rest of Roger's family and the families of the other thirteen who fell here. The sight of Wally's determined face is enough to bring me to tears and I am not ashamed of them and I don't try to hide them. We hike with a hundred others, and we're all close to tears here. It's not a time to be strong. It's a time to lean on others, and I do. I still have a hard time with letting people in, but I allow myself to be hugged, and hug back.

I sit at Roger's cross with the smokejumper and Steve and other firefighters from my past and we ceremoniously sip a flask in Roger's memory. We tell stories that make us howl with laughter. Stories of Roger, worse for wear from a party the night before, grabbing a radio and alerting the dispatchers for some reason known only to him: "We're surrounded by wolves!" Stories of him carefully sewing a tiny Nomex outfit for my stuffed bear. Wrestling matches gone wrong. There are more stories than I can remember, but I am glad. Each story means a memory, and if there are enough, it will be another twenty years before we run out.

A jump plane flies over and drops fourteen streamers, one for each of the fallen, and I watch them drop into the dense oak, grown up thick now, ready to burn again.

Last year nineteen young men died in Arizona in an eerily similar way, and I'm left wondering what we have all learned. It's hard to believe that we still allow this to happen, men and women marching into desolate places to test themselves against fire that is not threatening anything except for woods and brush. Why not let it burn, take our chances, accept that we live in a place that has always known fire?

I still have a complicated relationship with firefighting, like a bad romance I can't quite relinquish. Though I rarely get out anymore, there are still times when I catch a glimpse of something so pure and close to the bone that it reminds me of how it used to be, when I was young, when Roger and Jen and I walked through prairies together, lines of flame behind us. "Bring fire with you," Jed used to tell us, and that order always struck me as being beautiful and prophetic, the way I wanted to live my life always.

Families gather at each cross and I visit each one. I know a little bit of each story by now, and have met some of the relatives over the last twenty years. A tow-headed four-year-old boy toddles up the dusty, eroded trail, refusing a piggyback ride. There are other children and young adults who were not alive when this firestorm took the lives of people they will never know, but who will never be forgotten.

I decide to hike down alone, away from the crowd. Part of the trail follows the old fireline, and I imagine the sound of Roger's boots as he carried cubies of water down to the crew. I imagine the innocent blue of the sky, just another day, just another fire. All fires are the same until they aren't.

Roger will never know what he meant to me, because I never told him. Like the rest of the men and women I shared the fireline with, I assumed we would run into each other year after year, season after season. When I drove to my parents' house the spring before he died, I thought about taking the hundred-mile detour to see him and meet his parents before we all headed out west for the summer. In the end I decided not to, since we would surely see each other again in Florida, only six months away. Instead I met his parents for the first time over a coffin.

As I carefully descend the trail, I can feel the breath of nineteen others in line, a whole crew, all of us carrying our packs and our pulaskis and our big dreams. I can hear the sound of a helicopter carrying water and a lead plane scouting the route for an air tanker. I can still taste sweat and grit on my lips and the heat from an unseen fire. Just like Roger's memory, none of this will never really leave me.

I live now in a place that holds its secrets close, a town at the end of the road in Eastern Oregon. I moved here after Alaska, knowing that I wanted to find home. This country seemed like a good bet, remote and wild. It wraps itself up tight in prickly blackberry branches and oily poison ivy.

But home takes a while to find. *You need to live here longer,* this country seems to be telling me. *Four years and you expect everything from me? You need to earn it.* A lover of instant gratification, I am forced to take it slow. I search for the old trails braiding the landscape, trails that are melting into the wrinkled folds of the canyon an inch at a time. I look for the old homesteads, crumbling into dust and rotten boards. I piece it all together, a little bit at a time.

For years I made up my life as I went along. I had no use for plans. Slide in, suck the sweetness from a place, and move on. That is what I have always done. Everything I owned, everything worth keeping fit in the backseat of a Chevy Chevette. I was the girl driving barefoot, high on Vivarin and cinnamon gum, nothing slowing her down. Old lovers in the rearview, a stew of regret and anticipation in my heart, an endless clock ticking, move on, move on, don't stop. For years I followed the fire season from Alaska to Florida and everywhere in between. I rode the crest of a wave I thought would never end.

But everything ends. Roger died high on a mountain called Storm King, trapped in a fire he could not outrun. I was luckier a month later, on a different fire, in a different state, making it to a road in time. When I decided to trade in fire for someone I could love for more than just one season, for a dog, for a vacuum cleaner and a library card, I picked one of the most remote places I could find and still set my feet on pavement if I needed to. I wanted an escape route, just in case.

I now live in a wide valley with a river at its heart. Mountains form one boundary, a prairie the other. It's a long way from here to anywhere. The highway dead-ends here at a glacial lake. You cannot go any farther. Even though on a clear day I can see the Idaho side, it takes hours of bad road and crossing the Snake River to get there. It is a dead end, a sanctuary, a blessing, not the prison I once imagined.

Four years here and I feel only a little closer to belonging. I am still a hesitant searcher. I am slow to open my guarded heart. I let people in with a trickle, not a torrent. There have been too many years of moving, eleven states in twenty years, too much left behind. When you leave a place a door slams shut; the mountains move to fill the gap you have left. I have tried to go back to places I have left and there is no room there. I tell myself that I am done with moving, done with giving everything away to the thrift shop and starting blank-slated in some new town, some new state. Instead I plant trees that will take years to reach my shoulders. I seek permanence. I seek home.

This country, though. It tricks me over and over. Like a man you can't quite forget, it keeps me coming back for more, forgetting the desert-dry of my mouth and the desperate hope that there will be water in Somers Creek to save me one more time. I forget the rattle of a fat-bodied snake near my ankle. I forget the icy breath of a sudden April snowstorm and the hiss of lightning as I climb high to the canyon rim, exposed. I remember only this: the blush-colored kiss of last sun on the high rim, the endless silence.

When I got married for the first time, a brief and painful interlude when I was still fighting fire, my old friends in the seasonal tribe laughed and laughed. "The last of the great ones falls," they said. The marriage didn't last and neither did my resolve to stay put. The rest of my friends had all been married years ago, falling like bowling pins in a flurry of white dresses. They quit the road until it felt like I was the only one left out there.

This country has crept up on me without me noticing. Whatever you were before means nothing here. You can reinvent yourself, fall deep into the rhythms of a place ancient as time. I want to know what this old country knows. I want to feel its bones. I want to listen to the slow pulse of its heart through the canyon walls. I want to learn the slow river carving deep into stone. There seems to be a truth here that I cannot quite grasp, something real and honest and plain, something lasting, something I have been missing in my headlong flight. It might be community. It might be refuge. It might be hope.

When I used to drive across the country, safe in my cocoon of turning wheels, the lights from the little towns spread out like glowing embers on the Texas plains. I could see them for miles, each little spark a house with people inside. I used to feel sorry for those people. They were like birds that had lost their wings and did not remember flight, I thought. Far better to be me, a wind-tousled girl with no attachments, slipping easily from one skin to another.

Now I am one of those kinds of people, my light burning brightly against the darkness. The years have piled up like snow. I know just enough to believe that I might be here forever.

I won't lie. There are times I page through my tattered road atlas and think about trying it again. Surely I am not too old; surely I haven't forgotten how to lay down a line of fire in a prairie, how to dig a cup trench? I think about filling up the truck and heading west, or east, or south. I remember the road, a quirky and beautiful place. I saw strangers like me at the rest stops, their cars stuffed with bicycles and boxes. We were brothers and sisters traveling the major arteries of America, heading to the next fire season. To each of us, the road was as familiar as a neighborhood. It was a river, carrying us to freedom, away from anything that might want to tie us down. In the end, though, I take a breath and the thought passes. Another day goes by to stack up to forever.

This country breaks my heart, but only a little bit. It cracks my heart's solid core enough to know what the fuss is all about. Love and community and staying in place. Potlucks and fundraisers for people in trouble. Someone who will feed your dogs or humor you with a mindless slog through knee-deep snow just because. That quiet, good feeling you get, safe in your house on a night crusty with stars and new snow when you hear restless tires passing by, all of the cars, all of the cars with people inside them looking for home.

Every once in a while, though, fire slinks back into town. It has been gone from my life a good long time. Long enough for me almost to forget how it was. *Remember?* fire asks. *Remember how it was, back then?*

Of course I do. Who can forget lovers long gone? I miss the fire life sometimes. It was simple, uncomplicated, not like staying is. Staying means I face up to myself in the mirror instead of running away. If I do not like a job, or want out of a bad situation, I can't just hop in the car and leave it all behind. I have to let someone I love slip through the cracks in my armor.

There are plenty of men in this valley too, but I am gun-shy. I make no promises, even when I meet a blue-eyed skier with unruly hair he attempts to contain in a ponytail. I meet him in winter and I stand at the base of the ski hill, watching him drop in circles toward me. The way he moves on snow is the way I used to feel fighting fire, without boundaries. Free of constraint. Free from gravity. I like that he knows that feeling, even though he comes by it a different way. I think that because of this he can understand the life I have lived up until now.

When we ski together I am tentative, not wanting to commit to the rate of speed it takes to get down the hills. He skis in front of me and stops far below, holding out his arms.

"I'll catch you before you fall," he hollers.

I am not used to this. I'm used to going it alone and at first I try to ski past him, doubting. But at the last minute I turn and plow into him and he does, he catches me and we laugh as the snow falls around us. I like this, I discover. Leaning on someone does not have to be a weakness.

But I set the ground rules right away. No commitments, no promises. No marriage, ever again. He thinks about this on one of our summer backpack trips, squinting up at Traverse Ridge above us, mentally figuring out a ski line once snow falls. We climb up there, far above our tent. On the flat ridge we find perfect round balls of stone, striped yellow and orange. These used to be on the bottom of a river, he tells me, setting one in my palm. I turn it over and over in my hands, imagining its journey.

I peer down three thousand feet to where the river now runs. It has carved out this valley over hundreds of years, slipping down through stone, taking a little more each season.

Everything changes.

"I'll never get married again," I say, but not quite as firmly.

"Are you sure?" he asks.

We stand together in a lush green bowl. It is August, the full bloom of summer before a deep descent into winter. Shallow silver streams thread down its sides and wind lazily through the meadow. Our tent, pockmarked with dew, flaps slowly in a subtle breeze. His three dogs chase pikas, never catching them but always hoping, their white fur blowing in the wind.

I test him. I might apply for a seven-month writing residency, I tell him, one in a remote river corridor where the only communication is a radio telephone. I might go to Antarctica. I might go back to fire. I might leave, forever.

Unlike all the others, he does not beg me to stay. He does not begin a retreat of his own, closing down. He does not threaten me with the promise of other women. When he does speak, he is calm. His face is open, unworried. "I'll miss you terribly," he says. "But I'll do whatever I can to help."

Another night we sleep late in a place where I can hear the river, boulders loosened by snowmelt tumbling through the current. Sunlight falls on our faces. It's later than I planned to wake up.

There are trails to hike, things to do. "Do you know it's eight already?" I ask him.

He murmurs sleepily, not holding back one word or thought. "Do you know how much I love you?"

The way he asks me to marry him is this: we snowshoe up to a ski hut deep in the mountains, carrying our packs.

He takes a piece of gold thread and carefully cuts it with his Leatherman. He winds it around my finger.

It is perfect.

I had thought this was lost to me. I know now I was wrong. I say yes.

I won't lie about this either: there are stretches of time now, perhaps weeks, when I don't think about Roger anymore. He pops up in unexpected places, though: when I find his old ball cap as I clean out the shed, when I try to throw out my fire boots but place them, again, on the shelf. When this happens his memory is just as clear and sharp as it was twenty years ago. I can hear his infectious laugh and even the cough that wore on my last nerve. I can see him wading through sawgrass, singing the wrong lyrics to a song. I can see the three of us, Roger, Jen, and me, walking our parallel lines, trusting that the others are watching out for us, walking, always bringing fire with us.